ne of the World and Other Scientific Essays

John Joly

Alpha Editions

This edition published in 2024

ISBN : 9789367240793

Design and Setting By
Alpha Editions
www.alphaedis.com
Email - info@alphaedis.com

Contents

PREFACE

Tins volume contains twelve essays written at various times during recent years. Many of them are studies contributed to Scientific Reviews or delivered as popular lectures. Some are expositions of views the scientific basis of which may be regarded as established. Others—the greater number—may be described as attempting the solution of problems which cannot be approached by direct observation.

The essay on The Birth-time of the World is based on a lecture delivered before the Royal Dublin Society. The subject has attracted much attention within recent years. The age of the Earth is, indeed, of primary importance in our conception of the longevity of planetary systems. The essay deals with the evidence, derived from the investigation of purely terrestrial phenomena, as to the period which has elapsed since the ocean condensed upon the Earth's surface. Dr. Decker's recent addition to the subject appeared too late for inclusion in it. He finds that the movements (termed isostatic) which geologists recognise as taking place deep in the Earth's crust, indicate an age of the same order of magnitude

as that which is inferred from the statistics of denudative history.[1]

The subject of _Denudation_ naturally arises from the first essay. In thinking over the method of finding the age of the ocean by the accumulation of sodium therein, I perceived so long ago as 1899, when my first paper was published, that this method afforded a means of ascertaining the grand total of denudative work effected on the Earth's surface since the beginning of geological time; the resulting knowledge in no way involving any assumption as to the duration of the period comprising the denudative actions. This idea has been elaborated in various publications since then, both by myself and by others. "Denudation," while including a survey of the subject generally, is mainly a popular account of this method and its results. It closes with a reference to the fascinating problems presented by the inner nature of sedimentation: a branch of science to which I endeavoured to contribute some years ago.

Mountain Genesis first brings in the subject of the geological intervention of radioactivity. There can, I believe, be no doubt as to the influence of transforming elements upon the developments of the surface features of the Earth; and, if I am right, this source of thermal energy is mainly responsible for that local accumulation of wrinkling which we term mountain chains. The

[1] Bull. Geol. Soc. America, vol. xxvi, March 1915.

paper on _Alpine Structure_ is a reprint from "Radioactivity and Geology," which for the sake of completeness is here included. It is directed to the elucidation of a detail of mountain genesis: a detail which enters into recent theories of Alpine development. The weakness of the theory of the "horst" is manifest, however, in many of its other applications; if not, indeed, in all.

The foregoing essays on the physical influences affecting the surface features of the Earth are accompanied by one entitled _The Abundance of Life._ This originated amidst the overwhelming presentation of life which confronts us in the Swiss Alps. The subject is sufficiently inspiring. Can no fundamental reason be given for the urgency and aggressiveness of life? Vitality is an ever-extending phenomenon. It is plain that the great principles which have been enunciated in explanation of the origin of species do not really touch the problem. In the essay—which is an early one (1890)—the explanation of the whole great matter is sought—and as I believe found—in the attitude of the organism towards energy external to it; an attitude which results in its evasion of the retardative and dissipatory effects which prevail in lifeless dynamic systems of all kinds.

Other Minds than Ours? attempts a solution of the vexed question of the origin of the Martian "canals." The essay is an abridgment of two popular lectures on the subject. I had previously written an account of my views which carried the enquiry as far as it was in

my power to go. This paper appeared in the "Transactions of the Royal Dublin Society, 1897." The theory put forward is a purely physical one, and, if justified, the view that intelligent beings exist in Mars derives no support from his visible surface features; but is, in fact, confronted with fresh difficulties.

Pleochroic Haloes is a popular exposition of an inconspicuous but very beautiful phenomenon of the rocks. Minute darkened spheres—a microscopic detail—appear everywhere in certain of the rock minerals. What are they? The discoveries of recent radioactive research—chiefly due to Rutherford—give the answer. The measurements applied to the little objects render the explanation beyond question. They turn out to be a quite extraordinary record of radioactive energy; a record accumulated since remote geological times, and assuring us, indirectly, of the stability of the chemical elements in general since the beginning of the world. This assurance is, without proof, often assumed in our views on the geological history of the Globe.

Skating is a discourse, with a recent addition supporting the original thesis. It is an illustration of a common experience—the explanation of an unimportant action involving principles the most influential considered as a part of Nature's resources.

The address on _The Latent Image_ deals with a subject which had been approached by various writers before the time of my essay; but, so far as I know, an explanation

based on the facts of photo-electricity had not been attempted. Students of this subject will notice that the views expressed are similar to those subsequently put forward by Lenard and Saeland in explanation of phosphorescence. The whole matter is of more practical importance than appears at first sight, for the photoelectric nature of the effects involved in the radiative treatment of many cruel diseases seems to be beyond doubt.

It was in connection with photo-electric science that I was led to take an interest in the application of radioactivity in medicine. The lecture on _The Use of Radium in Medicine_ deals with this subject. Towards the conclusion of this essay reference will be found to a practical outcome of such studies which, by improving on the methods, and facilitating the application, of radioactive treatment, has, in the hands of skilled medical men, already resulted in the alleviation of suffering.

Leaving out much which might well appear in a prefatory notice, a word should yet be added respecting the illustrations of scenery. They are a small selection from a considerable number of photographs taken during my summer wanderings in the Alps in company with Henry H. Dixon. An exception is Plate X, which is by the late Dr. Edward Stapleton. From what has been said above, it will be gathered that these illustrations are fitly included among pages which owe so much to Alpine inspiration. They illustrate the subjects dealt with, and, it is to be hoped, they will in some cases recall to the reader scenes which have in past times influenced his thoughts in the same manner; scenes which in their endless perspective seem to reduce to their proper insignificance the lesser things of life.

My thanks are due to Mr. John Murray for kindly consenting to the reissue of the essay on _The Birth-time of the World_ from the pages of _Science Progress_; to Messrs. Constable & Co. for leave to reprint _Pleochroic Haloes_ from _Bedrock_, and also to make some extracts from _Radioactivity and Geology_; and to the Council of the Royal Dublin Society for permission to republish certain papers from the Proceedings of the Society.

Iveagh Geological Laboratory, Trinity College, Dublin.

July, 1915.

THE BIRTH-TIME OF THE WORLD [1]

LONG ago Lucretius wrote: "For lack of power to solve the question troubles the mind with doubts, whether there was ever a birth-time of the world and whether likewise there is to be any end." "And if" (he says in answer) "there was no birth-time of earth and heaven and they have been from everlasting, why before the Theban war and the destruction of Troy have not other poets as well sung other themes? Whither have so many deeds of men so often passed away, why live they nowhere embodied in lasting records of fame? The truth methinks is that the sum has but a recent date, and the nature of the world is new and has but lately had its commencement."[2]

Thus spake Lucretius nearly 2,000 years ago. Since then we have attained another standpoint and found very different limitations. To Lucretius the world commenced with man, and the answer he would give to his questions was in accord with his philosophy: he would date the birth-time of the world from the time when

[1] A lecture delivered before the Royal Dublin Society, February 6th, 1914. _Science Progress_, vol. ix., p. 37

[2] _De Rerum Natura_, translated by H. A. J. Munro (Cambridge, 1886).

poets first sang upon the earth. Modern Science has along with the theory that the Earth dated its beginning with the advent of man, swept utterly away this beautiful imagining. We can, indeed, find no beginning of the world. We trace back events and come to barriers which close our vista—barriers which, for all we know, may for ever close it. They stand like the gates of ivory and of horn; portals from which only dreams proceed; and Science cannot as yet say of this or that dream if it proceeds from the gate of horn or from that of ivory.

In short, of the Earth's origin we have no certain knowledge; nor can we assign any date to it. Possibly its formation was an event so gradual that the beginning was spread over immense periods. We can only trace the history back to certain events which may with considerable certainty be regarded as ushering in our geological era.

Notwithstanding our limitations, the date of the birth-time of our geological era is the most important date in Science. For in taking into our minds the spacious history of the universe, the world's age must play the part of time-unit upon which all our conceptions depend. If we date the geological history of the Earth by thousands of years, as did our forerunners, we must shape our ideas of planetary time accordingly; and the duration of our solar system, and of the heavens, becomes comparable with that of the dynasties of ancient nations. If by millions of years, the sun and stars are proportionately venerable. If by hundreds or thousands of millions of

years the human mind must consent to correspondingly vast epochs for the duration of material changes. The geological age plays the same part in our views of the duration of the universe as the Earth's orbital radius does in our views of the immensity of space. Lucretius knew nothing of our time-unit: his unit was the life of a man. So also he knew nothing of our space-unit, and he marvels that so small a body as the sun can shed so much, heat and light upon the Earth.

A study of the rocks shows us that the world was not always what it now is and long has been. We live in an epoch of denudation. The rains and frosts disintegrate the hills; and the rivers roll to the sea the finely divided particles into which they have been resolved; as well as the salts which have been leached from them. The sediments collect near the coasts of the continents; the dissolved matter mingles with the general ocean. The geologist has measured and mapped these deposits and traced them back into the past, layer by layer. He finds them ever the same; sandstones, slates, limestones, etc. But one thing is not the same. _Life_ grows ever less diversified in character as the sediments are traced downwards. Mammals and birds, reptiles, amphibians, fishes, die out successively in the past; and barren sediments ultimately succeed, leaving the first beginnings of life undecipherable. Beneath these barren sediments lie rocks collectively differing in character from those above: mainly volcanic or poured out from fissures in

the early crust of the Earth. Sediments are scarce among these materials.[1]

There can be little doubt that in this underlying floor of igneous and metamorphic rocks we have reached those surface materials of the earth which existed before the long epoch of sedimentation began, and before the seas came into being. They formed the floor of a vaporised ocean upon which the waters condensed here and there from the hot and heavy atmosphere. Such were the probable conditions which preceded the birth-time of the ocean and of our era of life and its evolution.

It is from this epoch we date our geological age. Our next purpose is to consider how long ago, measured in years, that birth-time was.

That the geological age of the Earth is very great appears from what we have already reviewed. The sediments of the past are many miles in collective thickness: yet the feeble silt of the rivers built them all from base to summit. They have been uplifted from the seas and piled into mountains by movements so slow that during all the time man has been upon the Earth but little change would have been visible. The mountains have again been worn down into the ocean by denudation and again younger mountains built out of their redeposited materials. The contemplation of such vast events

[1] For a description of these early rocks, see especially the monograph of Van Hise and Leith on the pre-Cambrian Geology of North America (Bulletin 360, U.S. Geol. Survey).

prepares our minds to accept many scores of millions of years or hundreds of millions of years, if such be yielded by our calculations.

THE AGE AS INFERRED FROM THE THICKNESS OF THE SEDIMENTS

The earliest recognised method of arriving at an estimate of the Earth's geological age is based upon the measurement of the collective sediments of geological periods. The method has undergone much revision from time to time. Let us briefly review it on the latest data.

The method consists in measuring the depths of all the successive sedimentary deposits where these are best developed. We go all over the explored world, recognising the successive deposits by their fossils and by their stratigraphical relations, measuring their thickness and selecting as part of the data required those beds which we believe to most completely represent each formation. The total of these measurements would tell us the age of the Earth if their tale was indeed complete, and if we knew the average rate at which they have been deposited. We soon, however, find difficulties in arriving at the quantities we require. Thus it is not easy to measure the real thickness of a deposit. It may be folded back upon itself, and so we may measure it twice over. We may exaggerate its thickness by measuring it not quite straight across the bedding or by unwittingly including volcanic materials. On the other hand, there

may be deposits which are inaccessible to us; or, again, an entire absence of deposits; either because not laid down in the areas we examine, or, if laid down, again washed into the sea. These sources of error in part neutralise one another. Some make our resulting age too long, others make it out too short. But we do not know if a balance of error does not still remain. Here, however, is a table of deposits which summarises a great deal of our knowledge of the thickness of the stratigraphical accumulations. It is due to Sollas.[1]

	Feet.
Recent and Pleistocene --	4,000
Pliocene --	13,000
Miocene --	14,000
Oligocene --	2,000
Eocene --	20,000
	63,000
Upper Cretaceous --	24,000
Lower Cretaceous --	20,000
Jurassic --	8,000
Trias --	7,000
	69,000
Permian --	2,000
Carboniferous --	29,000
Devonian --	22,000
	63,000
Silurian --	15,000

Ordovician	--	17,000
Cambrian	--	6,000
		58,000

Algonkian—Keeweenawan	--	50,000
Algonkian—Animikian	--	14,000
Algonkian—Huronian	--	18,000
		82,000

Archæan	--	?
Total	--	335,000 feet.

[1] Address to the Geol. Soc. of London, 1509.

In the next place we require to know the average rate at which these rocks were laid down. This is really the weakest link in the chain. The most diverse results have been arrived at, which space does not permit us to consider. The value required is most difficult to determine, for it is different for the different classes of material, and varies from river to river according to the conditions of discharge to the sea. We may probably take it as between two and six inches in a century.

Now the total depth of the sediments as we see is about 335,000 feet (or 64 miles), and if we take the rate of collecting as three inches in a hundred years we get the time for all to collect as 134 millions of years. If the rate be four inches, the time is soo millions of years, which is the figure Geikie favoured, although his result was based on somewhat different data. Sollas most recently finds 80 millions of years.[1]

THE AGE AS INFERRED FROM THE MASS OF THE SEDIMENTS

In the above method we obtain our result by the measurement of the linear dimensions of the sediments. These measurements, as we have seen, are difficult to arrive at. We may, however, proceed by measurements of the mass of the sediments, and then the method becomes more definite. The new method is pursued as follows:

[1] Geikie, _Text Book of Geology_ (Macmillan, 1903), vol. i., p. 73, _et seq._ Sollas, _loc. cit._ Joly, _Radioactivity and Geology_ (Constable, 1909), and Phil. Mag., Sept. 1911.

The total mass of the sediments formed since denudation began may be ascertained with comparative accuracy by a study of the chemical composition of the waters of the ocean. The salts in the ocean are undoubtedly derived from the rocks; increasing age by age as the latter are degraded from their original character under the action of the weather, etc., and converted to the sedimentary form. By comparing the average chemical composition of these two classes of material—the primary or igneous rocks and the sedimentary—it is easy to arrive at a knowledge of how much of this or that constituent was given to the ocean by each ton of primary rock which was denuded to the sedimentary form. This, however, will not assist us to our object unless the ocean has retained the salts shed into it. It has not generally done so. In the case of every substance but one the ocean continually gives up again more or less of the salts supplied to it by the rivers. The one exception is the element sodium. The great solubility of its salts has protected it from abstraction, and it has gone on collecting during geological time, practically in its entirety. This gives us the clue to the denudative history of the Earth.[1]

The process is now simple. We estimate by chemical examination of igneous and sedimentary rocks the amount of sodium which has been supplied to the ocean per ton of sediment produced by denudation. We also calculate

[1] _Trans. R.D.S._, May, 1899.

the amount of sodium contained in the ocean. We divide the one into the other (stated, of course, in the same units of mass), and the quotient gives us the number of tons of sediment. The most recent estimate of the sediments made in this manner affords 56 x 10^{16} tonnes.[1]

Now we are assured that all this sediment was transported by the rivers to the sea during geological time. Thus it follows that, if we can estimate the average annual rate of the river supply of sediments to the ocean over the past, we can calculate the required age. The land surface is at present largely covered with the sedimentary rocks themselves. Sediment derived from these rocks must be regarded as, for the most part, purely cyclical; that is, circulating from the sea to the land and back again. It does not go to increase the great body of detrital deposits. We cannot, therefore, take the present river supply of sediment as representing that obtaining over the long past. If the land was all covered still with primary rocks we might do so. It has been estimated that about 25 per cent. of the existing continental area is covered with archæan and igneous rocks, the remainder being sediments.[2] On this estimate we may find valuable

[1] Clarke, _A Preliminary Study of Chemical Denudation_ (Washington, 1910). My own estimate in 1899 (_loc. cit._) made as a test of yet another method of finding the age, showed that the sediments may be taken as sufficient to form a layer 1.1 mile deep if spread uniformly over the continents; and would amount to 64 x 10^{18} tons.

[2] Van Tillo, _Comptes Rendues_ (Paris), vol. cxiv., 1892.

major and minor limits to the geological age. If we take 25 per cent. only of the present river supply of sediment, we evidently fix a major limit to the age, for it is certain that over the past there must have been on the average a faster supply. If we take the entire river supply, on similar reasoning we have what is undoubtedly a minor limit to the age.

The river supply of detrital sediment has not been very extensively investigated, although the quantities involved may be found with comparative ease and accuracy. The following table embodies the results obtained for some of the leading rivers.[1]

	Mean annual discharge in cubic feet per second.	Total annual sediment in thousands of tons.	Ratio of sediment to water by weight.
Potomac -	20,160	5,557	1 : 3.575
Mississippi -	610,000	406,250	1 : 1,500
Rio Grande -	1,700	3,830	1 : 291
Uruguay -	150,000	14,782	1 : 10,000
Rhone -	65,850	36,000	1 : 1,775
Po -	62,200	67,000	1 : 900
Danube -	315,200	108,000	1 : 2,880
Nile -	113,000	54,000	1 : 2,050
Irrawaddy -	475,000	291,430	1 : 1,610
Mean -	201,468	109,650	1 : 2,731

We see that the ratio of the weight of water to the

[1] Russell, _River Development_ (John Murray, 1888).

weight of transported sediment in six out of the nine rivers does not vary widely. The mean is 2,730 to 1. But this is not the required average. The water-discharge of each river has to be taken into account. If we ascribe to the ratio given for each river the weight proper to the amount of water it discharges, the proportion of weight of water to weight of sediment, for the whole quantity of water involved, comes out as 2,520 to 1.

Now if this proportion holds for all the rivers of the world—which collectively discharge about 27 x 1012 tonnes of water per annum—the river-born detritus is 1.07 x 1010 tonnes. To this an addition of 11 per cent. has to be made for silt pushed along the river-bed.[1] On these figures the minor limit to the age comes out as 47 millions of years, and the major limit as 188 millions. We are here going on rather deficient estimates, the rivers involved representing only some 6 per cent. of the total river supply of water to the ocean. But the result is probably not very far out.

We may arrive at a probable age lying between the major and minor limits. If, first, we take the arithmetic mean of these limits, we get 117 millions of years. Now this is almost certainly excessive, for we here assume that the rate of covering of the primary rocks by sediments was uniform. It would not be so, however, for the rate of supply of original sediment must have been continually diminishing

[1] According to observations made on the Mississippi (Russell, _loc. cit._).

during geological time, and hence we may assume that the rate of advance of the sediments on the primary rocks has also been diminishing. Now we may probably take, as a fair assumption, that the sediment-covered area was at any instant increasing at a rate proportionate to the rate of supply of sediment; that is, to the area of primary rocks then exposed. On this assumption the age is found to be 87 millions of years.

THE AGE BY THE SODIUM OF THE OCEAN

I have next to lay before you a quite different method. I have already touched upon the chemistry of the ocean, and on the remarkable fact that the sodium contained in it has been preserved, practically, in its entirety from the beginning of geological time.

That the sea is one of the most beautiful and magnificent sights in Nature, all admit. But, I think, to those who know its story its beauty and magnificence are ten-fold increased. Its saltness it due to no magic mill. It is the dissolved rocks of the Earth which give it at once its brine, its strength, and its buoyancy. The rivers which we say flow with "fresh" water to the sea nevertheless contain those traces of salt which, collected over the long ages, occasion the saltness of the ocean. Each gallon of river water contributes to the final result; and this has been going on since the beginning of our era. The mighty total of the rivers is 6,500 cubic miles of water in the year!

There is little doubt that the primeval ocean was in the condition of a fresh-water lake. It can be shown that a primitive and more rapid solution of the original crust of the Earth by the slowly cooling ocean would have given rise to relatively small salinity. The fact is, the quantity of salts in the ocean is enormous. We are only now concerned with the sodium; but if we could extract all the rock-salt (the chloride of sodium) from the ocean we should have enough to cover the entire dry land of the Earth to a depth of 400 feet. It is this gigantic quantity which is going to enter into our estimate of the Earth's age. The calculated mass of sodium contained in this rock-salt is 14,130 million million tonnes.

If now we can determine the rate at which the rivers supply sodium to the ocean, we can determine the age.[1] As the result of many thousands of river analyses, the total amount of sodium annually discharged to the ocean

[1] _Trans. R.D.S._, 1899. A paper by Edmund Halley, the astronomer, in the _Philosophical Transactions of the Royal Society_ for 1715, contains a suggestion for finding the age of the world by the following procedure. He proposes to make observations on the saltness of the seas and ocean at intervals of one or more centuries, and from the increment of saltness arrive at their age. The measurements, as a matter of fact, are impracticable. The salinity would only gain (if all remained in solution) one millionth part in Too years; and, of course, the continuous rejection of salts by the ocean would invalidate the method. The last objection also invalidates the calculation by T.

Mellard Reade (_Proc. Liverpool Geol. Soc._, 1876) of a minor limit to the age by the calcium sulphate in the ocean. Both papers were quite unknown to me when working out my method. Halley's paper was, I think, only brought to light in 1908.

by all the rivers of the world is found to be probably not far from 175 million tonnes.[1] Dividing this into the mass of oceanic sodium we get the age as 80.7 millions of years. Certain corrections have to be applied to this figure which result in raising it to a little over 90 millions of years. Sollas, as the result of a careful review of the data, gets the age as between 80 and 150 millions of years. My own result[2] was between 80 and 90 millions of years; but I subsequently found that upon certain extreme assumptions a maximum age might be arrived at of 105 millions of years.[3] Clarke regards the 80.7 millions of years as certainly a maximum in the light of certain calculations by Becker.[4]

The order of magnitude of these results cannot be shaken unless on the assumption that there is something entirely misleading in the existing rate of solvent denudation. On the strength of the results of another and

[1] F. W. Clarke, _A Preliminary Study of Chemical Denudation_ (Smithsonian Miscellaneous Collections, 1910).

[2] _Loc. cit._

[3] "The Circulation of Salt and Geological Time" (Geol. Mag., 1901, p. 350).

[4] Becker (loc. cit.), assuming that the exposed igneous and archæan rocks alone are responsible for the supply of sodium to the ocean, arrives at 74 millions of years as the geological age. This matter was discussed by me formerly (Trans. R.D.S., 1899, pp. 54 _et seq._). The assumption made is, I believe, inadmissible. It is not supported by river analyses, or by the chemical character of residual soils from sedimentary rocks. There may be some convergence in the rate of solvent denudation, but—as I think on the evidence—in our time unimportant.

entirely different method of approaching the question of the Earth's age (which shall be presently referred to), it has been contended that it is too low. It is even asserted that it is from nine to fourteen times too low. We have then to consider whether such an enormous error can enter into the method. The measurements involved cannot be seriously impugned. Corrections for possible errors applied to the quantities entering into this method have been considered by various writers. My own original corrections have been generally confirmed. I think the only point left open for discussion is the principle of uniformitarianism involved in this method and in the methods previously discussed.

In order to appreciate the force of the evidence for uniformity in the geological history of the Earth, it is, of course, necessary to possess some acquaintance with geological science. Some of the most eminent geologists, among whom Lyell and Geikie[1] may be mentioned, have upheld the doctrine of uniformity. It must here suffice to dwell upon a few points having special reference to the matter under discussion.

The mere extent of the land surface does not, within limits, affect the question of the rate of denudation. This arises from the fact that the rain supply is quite insufficient to denude the whole existing land surface. About 30 per cent. of it does not, in fact, drain to the

[1] See especially Geikie's Address to Sect. C., Brit. Assoc. Rep., 1399.

ocean. If the continents become invaded by a great transgression of the ocean, this "rainless" area diminishes: and the denuded area advances inwards without diminution. If the ocean recedes from the present strand lines, the "rainless" area advances outwards, but, the rain supply being sensibly constant, no change in the river supply of salts is to be expected.

Age-long submergence of the entire land, or of any very large proportion of what now exists, is negatived by the continuous sequence of vast areas of sediment in every geologic age from the earliest times. Now sediment-receiving areas always are but a small fraction of those exposed areas whence the sediments are supplied.[1] Hence in the continuous records of the sediments we have assurance of the continuous exposure of the continents above the ocean surface. The doctrine of the permanency of the continents has in its main features been accepted by the most eminent authorities. As to the actual amount of land which was exposed during past times to denudative effects, no data exist to show it was very different from what is now exposed. It has been estimated that the average area of the North American continent over geologic time was about eight-tenths of its existing area.[2] Restorations of other continents, so far as they have been attempted, would not

[1] On the strength of the Mississippi measurements about 1 to 18 (Magee, _Am. Jour. of Sc._, 1892, p. 188).

[2] Schuchert, _Bull. Geol. Soc. Am._, vol. xx., 1910.

16

suggest any more serious divergency one way or the other.

That climate in the oceans and upon the land was throughout much as it is now, the continuous chain of teeming life and the sensitive temperature limits of protoplasmic existence are sufficient evidence.[1] The influence at once of climate and of elevation of the land may be appraised at their true value by the ascertained facts of solvent denudation, as the following table shows.

	Tonnes removed in solution per square mile per annum.	Mean elevation. Metres.
North America -	79	700
South America -	50	650
Europe -	100	300
Asia -	84	950
Africa -	44	650

In this table the estimated number of tonnes of matter in solution, which for every square mile of area the rivers convey to the ocean in one year, is given in the first column. These results are compiled by Clarke from a very large number of analyses of river waters. The second column of the table gives the mean heights in metres above sea level of the several continents, as cited by Arrhenius.[2]

Of all the denudation results given in the table, those relating to North America and to Europe are far the

[1] See also Poulton, Address to Sect. D., Brit. Assoc. Rep., 1896.

[2] _Lehybuch dev Kosmischen Physik_, vol. i., p. 347.

most reliable. Indeed these may be described as highly reliable, being founded on some thousands of analyses, many of which have been systematically pursued through every season of the year.

These show that Europe with a mean altitude of less than half that of North America sheds to the ocean 25 per cent. more salts. A result which is to be expected when the more important factors of solvent denudation are given intelligent consideration and we discriminate between conditions favouring solvent and detrital denudation respectively: conditions in many cases antagonistic.[1] Hence if it is true, as has been stated, that we now live in a period of exceptionally high continental elevation, we must infer that the average supply of salts to the ocean by the rivers of the world is less than over the long past, and that, therefore, our estimate of the age of the Earth as already given is excessive.

There is, however, one condition which will operate to unduly diminish our estimate of geologic time, and it is a condition which may possibly obtain at the present time. If the land is, on the whole, now sinking relatively to the ocean level, the denudation area tends, as we have seen, to move inwards. It will thus encroach upon regions which have not for long periods drained to the ocean. On such areas there is an accumulation of soluble salts which the deficient rivers have not been able to carry to the ocean. Thus the salt content of certain of

[1] See the essay on Denudation.

the rivers draining to the ocean will be influenced not only by present denudative effects, but also by the stored results of past effects. Certain rivers appear to reveal this unduly increased salt supply those which flow through comparatively arid areas. However, the flowoff of such tributaries is relatively small and the final effects on the great rivers apparently unimportant—a result which might have been anticipated when the extremely slow rate of the land movements is taken into account.

The difficulty of effecting any reconciliation of the methods already described and that now to be given increases the interest both of the former and the latter.

THE AGE BY RADIOACTIVE TRANSFORMATIONS

Rutherford suggested in 1905 that as helium was continually being evolved at a uniform rate by radioactive substances (in the form of the alpha rays) a determination of the age of minerals containing the radioactive elements might be made by measurements of the amount of the stored helium and of the radioactive elements giving rise to it, The parent radioactive substances are—according to present knowledge—uranium and thorium. An estimate of the amounts of these elements present enables the rate of production of the helium to be calculated. Rutherford shortly afterwards found by this method an age of 240 millions of years for a radioactive mineral of presumably remote age. Strutt, who carried

his measurements to a wonderful degree of refinement, found the following ages for mineral substances originating in different geological ages:

Oligocene - 8.4 millions of years.
Eocene - 31 millions of years.
Lower Carboniferous - 150 millions of years.
Archæan - 750 millions of years.

Periods of time much less than, and very inconsistent with, these were also found. The lower results are, however, easily explained if we assume that the helium—which is a gas under prevailing conditions—escapes in many cases slowly from the mineral.

Another product of radioactive origin is lead. The suggestion that this substance might be made available to determine the age of the Earth also originated with Rutherford. We are at least assured that this element cannot escape by gaseous diffusion from the minerals. Boltwood's results on the amount of lead contained in minerals of various ages, taken in conjunction with the amount of uranium or parent substance present, afforded ages rising to 1,640 millions of years for archæan and 1,200 millions for Algonkian time. Becker, applying the same method, obtained results rising to quite incredible periods: from 1,671 to 11,470 millions of years. Becker maintained that original lead rendered the determinations indefinite. The more recent results of Mr. A. Holmes support the conclusion that "original" lead may be present and may completely falsify results derived

from minerals of low radioactivity in which the derived lead would be small in amount. By rejecting such results as appeared to be of this character, he arrives at 370 millions of years as the age of the Devonian.

I must now describe a very recent method of estimating the age of the Earth. There are, in certain rock-forming minerals, colour-changes set up by radioactive causes. The minute and curious marks so produced are known as haloes; for they surround, in ringlike forms, minute particles of included substances which contain radioactive elements. It is now well known how these haloes are formed. The particle in the centre of the halo contains uranium or thorium, and, necessarily, along with the parent substance, the various elements derived from it. In the process of transformation giving rise to these several derived substances, atoms of helium—the alpha rays—projected with great velocity into the surrounding mineral, occasion the colour changes referred to. These changes are limited to the distance to which the alpha rays penetrate; hence the halo is a spherical volume surrounding the central substance.[1]

The time required to form a halo could be found if on the one hand we could ascertain the number of alpha rays ejected from the nucleus of the halo in, say, one year, and, on the other, if we determined by experiment just how many alpha rays were required to produce the same

[1] _Phil. Mag._, March, 1907 and February, 1910; also _Bedrock_, January, 1913. See _Pleochroic Haloes_ in this volume.

amount of colour alteration as we perceive to extend around the nucleus.

The latter estimate is fairly easily and surely made. But to know the number of rays leaving the central particle in unit time we require to know the quantity of radioactive material in the nucleus. This cannot be directly determined. We can only, from known results obtained with larger specimens of just such a mineral substance as composes the nucleus, guess at the amount of uranium, or it may be thorium, which may be present.

This method has been applied to the uranium haloes of the mica of County Carlow.[1] Results for the age of the halo of from 20 to 400 millions of years have been obtained. This mica was probably formed in the granite of Leinster in late Silurian or in Devonian times.

The higher results are probably the least in error, upon the data involved; for the assumption made as to the amount of uranium in the nuclei of the haloes was such as to render the higher results the more reliable.

This method is, of course, a radioactive method, and similar to the method by helium storage, save that it is free of the risk of error by escape of the helium, the effects of which are, as it were, registered at the moment of its production, so that its subsequent escape is of no moment.

[1] Joly and Rutherford, _Phil. Mag._, April, 1913.

REVIEW OF THE RESULTS

We shall now briefly review the results on the geological age of the Earth.

By methods based on the approximate uniformity of denudative effects in the past, a period of the order of 100 millions of years has been obtained as the duration of our geological age; and consistently whether we accept for measurement the sediments or the dissolved sodium. We can give reasons why these measurements might afford too great an age, but we can find absolutely no good reason why they should give one much too low.

By measuring radioactive products ages have been found which, while they vary widely among themselves, yet claim to possess accuracy in their superior limits, and exceed those derived from denudation from nine to fourteen times.

In this difficulty let us consider the claims of the radioactive method in any of its forms. In order to be trustworthy it must be true; (1) that the rate of transformation now shown by the parent substance has obtained throughout the entire past, and (2) that there were no other radioactive substances, either now or formerly existing, except uranium, which gave rise to lead. As regards methods based on the production of helium, what we have to say will largely apply to it also. If some unknown source of these elements exists we, of course, on our assumption overestimate the age.

As regards the first point: In ascribing a constant rate of change to the parent substance—which Becker (loc. cit.) describes as "a simple though tremendous extrapolation"—we reason upon analogy with the constant rate of decay observed in the derived radioactive bodies. If uranium and thorium are really primary elements, however, the analogy relied on may be misleading; at least, it is obviously incomplete. It is incomplete in a particular which may be very important: the mode of origin of these parent bodies—whatever it may have been—is different to that of the secondary elements with which we compare them. A convergence in their rate of transformation is not impossible, or even improbable, so far as we known.

As regards the second point: It is assumed that uranium alone of the elements in radioactive minerals is ultimately transformed to lead by radioactive changes. We must consider this assumption.

Recent advances in the chemistry of the radioactive elements has brought out evidence that all three lines of radioactive descent known to us—_i.e._ those beginning with uranium, with thorium, and with actinium—alike converge to lead.[1] There are difficulties in the way of believing that all the lead-like atoms so produced ("isotopes" of lead, as Soddy proposes to call them) actually remain as stable lead in the minerals. For one

[1] See Soddy's _Chemistry of the Radioactive Elements_ (Longmans, Green & Co.).

thing there is sometimes, along with very large amounts of thorium, an almost entire absence of lead in thorianites and thorites. And in some urano—thorites the lead may be noticed to follow the uranium in approximate proportionality, notwithstanding the presence of large amounts of thorium.[1] This is in favour of the assumption that all the lead present is derived from the uranium. The actinium is present in negligibly small amounts.

On the other hand, there is evidence arising from the atomic weight of lead which seems to involve some other parent than uranium. Soddy, in the work referred to, points this out. The atomic weight of radium is well known, and uranium in its descent has to change to this element. The loss of mass between radium and uranium-derived lead can be accurately estimated by the number of alpha rays given off. From this we get the atomic weight of uranium-derived lead as closely 206. Now the best determinations of the atomic weight of normal lead assign to this element an atomic weight of closely

[1] It seems very difficult at present to suggest an end product for thorium, unless we assume that, by loss of electrons, thorium E, or thorium lead, reverts to a substance chemically identical with thorium itself. Such a change—whether considered from the point of view of the periodic law or of the radioactive theory would involve many interesting consequences. It is, of course, quite possible that the nature of the conditions attending the deposition of the uranium ores, many of which are comparatively recent, are responsible for the difficulties observed. The thorium and uranium ores are, again, specially prone to alteration.

207. By a somewhat similar calculation it is deduced that thorium-derived lead would possess the atomic weight of 208. Thus normal lead might be an admixture of uranium- and thorium-derived

lead. However, as we have seen, the view that thorium gives rise to stable lead is beset with some difficulties.

If we are going upon reliable facts and figures, we must, then, assume: (a) That some other element than uranium, and genetically connected with it (probably as parent substance), gives rise, or formerly gave rise, to lead of heavier atomic weight than normal lead. It may be observed respecting this theory that there is some support for the view that a parent substance both to uranium and thorium has existed or possibly exists. The evidence is found in the proportionality frequently observed between the amounts of thorium and uranium in the primary rocks.[1] Or: (b) We may meet the difficulties in a simpler way, which may be stated as follows: If we assume that all stable lead is derived from uranium, and at the same time recognise that lead is not perfectly homogeneous in atomic weight, we must, of necessity, ascribe to uranium a similar want of homogeneity; heavy atoms of uranium giving rise to heavy

[1] Compare results for the thorium content of such rocks (appearing in a paper by the author Cong. Int. _de Radiologie et d'Electricité_, vol. i., 1910, p. 373), and those for the radium content, as collected in _Phil. Mag._, October, 1912, p. 697. Also A. L. Fletcher, _Phil. Mag._, July, 1910; January, 1911, and June, 1911. J. H. J. Poole, _Phil. Mag._, April, 1915

atoms of lead and light atoms of uranium generating light atoms of lead. This assumption seems to be involved in the figures upon, which we are going. Still relying on these figures, we find, however, that existing uranium cannot give rise to lead of normal atomic weight. We can only conclude that the heavier atoms of uranium have decayed more rapidly than the lighter ones. In this connection it is of interest to note the complexity of uranium as recently established by Geiger, although in this case it is assumed that the shorter-lived isotope bears the relation of offspring to the longer-lived and largely preponderating constituent. However, there does not seem to be any direct proof of this as yet.

From these considerations it would seem that unless the atomic weight of lead in uraninites, etc., is 206, the former complexity and more accelerated decay of uranium are indicated in the data respecting the atomic weights of radium and lead[1]. As an alternative view, we may assume, as in our first hypothesis, that some elementally different but genetically connected substance, decaying along branching lines of descent at a rate sufficient to practically remove the whole of it during geological time, formerly existed. Whichever hypothesis we adopt

[1] Later investigation has shown that the atomic weight of lead in uranium-bearing ores is about 206.6 (see Richards and Lembert, _Journ. of Am. Claem. Soc._, July, 1914). This result gives support to the view expressed above.

we are confronted by probabilities which invalidate time-measurements based on the lead and helium ratio in minerals. We have, in short, grave reason to question the measure of uniformitarianism postulated in finding the age by any of the known radioactive methods.

That we have much to learn respecting our assumptions, whether we pursue the geological or the radioactive methods of approaching the age of our era, is, indeed, probable. Whatever the issue it is certain that the reconciling facts will leave us with much more light than we at present possess either as respects the Earth's history or the history of the radioactive elements. With this necessary admission we leave our study of the Birth-Time of the World.

It has led us a long way from Lucretius. We do not ask if other Iliads have perished; or if poets before Homer have vainly sung, becoming a prey to all-consuming time. We move in a greater history, the landmarks of which are not the birth and death of kings and poets, but of species, genera, orders. And we set out these organic events not according to the passing generations of man, but over scores or hundreds of millions of years.

How much Lucretius has lost, and how much we have gained, is bound up with the question of the intrinsic value of knowledge and great ideas. Let us appraise knowledge as we would the Homeric poems, as some-

thing which ennobles life and makes it happier. Well, then, we are, as I think, in possession today of some of those lost Iliads and Odysseys for which Lucretius looked in vain.[1]

[1] The duration in the past of Solar heat is necessarily bound up with the geological age. There is no known means (outside speculative science) of accounting for more than about 30 million years of the existing solar temperature in the past. In this direction the age seems certainly limited to 100 million years. See a review of the question by Dr. Lindemann in Nature, April 5th, 1915.

DENUDATION

THE subject of denudation is at once one of the most interesting and one of the most complicated with which the geologist has to deal. While its great results are apparent even to the most casual observer, the factors which have led to these results are in many cases so indeterminate, and in some cases apparently so variable in influence, that thoughtful writers have even claimed precisely opposite effects as originating from, the same cause. Indeed, it is almost impossible to deal with the subject without entering upon controversial matters. In the following pages I shall endeavour to keep to broad issues which are, at the present day, either conceded by the greater number of authorities on the subject, or are, from their strictly quantitative character, not open to controversy.

It is evident, in the first place, that denudation—or the wearing away of the land surfaces of the earth—is mainly a result of the circulation of water from the ocean to the land, and back again to the ocean. An action entirely conditioned by solar heat, and without which it would completely cease and further change upon the land come to an end.

To what actions, then, is so great a potency of the

circulating water to be traced? Broadly speaking, we may classify them as mechanical and chemical. The first involves the separation of rock masses into smaller fragments of all sizes, down to the finest dust. The second involves the actual solution in the water of the rock constituents, which may be regarded as the final act of disintegration. The rivers bear the burden both of the comminuted and the dissolved materials to the sea. The mud and sand carried by their currents, or gradually pushed along their beds, represent the former; the invisible dissolved matter, only to be demonstrated to the eye by evaporation of the water or by chemical precipitation, represents the latter.

The results of these actions, integrated over geological time, are enormous. The entire bulk of the sedimentary rocks, such as sandstones, slates, shales, conglomerates, limestones, etc., and the salt content of the ocean, are due to the combined activity of mechanical and solvent denudation. We shall, later on, make an estimate of the magnitude of the quantities actually involved.

In the Swiss valleys we see torrents of muddy water hurrying along, and if we follow them up, we trace them to glaciers high among the mountains. From beneath the foot of the glacier, we find, the torrent has birth. The first debris given to the river is derived from the wearing of the rocky bed along which the glacier moves. The river of ice bequeaths to the river of water—of which it is the parent—the spoils which it has won from the rocks

The work of mechanical disintegration is, however, not restricted to the glacier's bed. It proceeds everywhere over the surface of the rocks. It is aided by the most diverse actions. For instance, the freezing and expansion of water in the chinks and cracks in those alpine heights where between sunrise and sunset the heat of summer reigns, and between sunset and sunrise the cold of winter. Again, under these conditions the mere change of surface temperature from night to day severely stresses the surface layers of the rocks, and, on the same principles as we explain the fracture of an unequally heated glass vessel, the rocks cleave off in slabs which slip down the steeps of the mountain and collect as screes in the valley. At lower levels the expansive force of vegetable growth is not unimportant, as all will admit who have seen the strong roots of the pines penetrating the crannies of the rocks. Nor does the river which flows in the bed of the valley act as a carrier only. Listening carefully we may detect beneath the roar of the alpine torrent the crunching and knocking of descending boulders. And in the potholes scooped by its whirling waters we recognise the abrasive action of the suspended sand upon the river bed.

A view from an Alpine summit reveals a scene of remarkable desolation (Pl. V, p. 40). Screes lie piled against the steep slopes. Cliffs stand shattered and ready to fall in ruins. And here the forces at work readily reveal themselves. An occasional wreath of white smoke among

the far-off peaks, followed by a rumbling reverberation, marks the fall of an avalanche. Water everywhere trickles through the shaly _débris_ scattered around. In the full sunshine the rocks are almost too hot to bear touching. A few hours later the cold is deadly, and all becomes a frozen silence. In such scenes of desolation and destruction, detrital sediments are actively being generated. As we descend into the valley we hear the deep voice of the torrents which are continually hurrying the disintegrated rocks to the ocean.

A remarkable demonstration of the activity of mechanical denudation is shown by the phenomenon of "earth pillars." The photograph (Pl. IV.) of the earth pillars of the Val d'Hérens (Switzerland) shows the peculiar appearance these objects present. They arise under conditions where large stones or boulders are scattered in a deep deposit of clay, and where much of the denudation is due to water scour. The large boulders not only act as shelter against rain, but they bind and consolidate by their mere weight the clay upon which they rest. Hence the materials underlying the boulders become more resistant, and as the surrounding clays are gradually washed away and carried to the streams, these compacted parts persist, and, finally, stand like walls or pillars above the general level. After a time the great boulders fall off and the underlying clay becomes worn by the rainwash to fantastic spikes and ridges. In the Val d'Hérens the earth pillars are formed

of the deep moraine stuff which thickly overlies the slopes of the valley. The wall of pillars runs across the axis of the valley, down the slope of the hill, and crosses the road, so that it has to be tunnelled to permit the passage of traffic. It is not improbable that some additional influence—possibly the presence of lime—has hardened the material forming the pillars, and tended to their preservation.

Denudation has, however, other methods of work than purely mechanical; methods more noiseless and gentle, but not less effective, as the victories of peace ate no less than those of war.

Over the immense tracts of the continents chemical work proceeds relentlessly. The rock in general, more especially the primary igneous rock, is not stable in presence of the atmosphere and of water. Some of the minerals, such as certain silicates and carbonates, dissolve relatively fast, others with extreme slowness. In the process of solution chemical actions are involved; oxidation in presence of the free oxygen of the atmosphere; attack by the feeble acid arising from the solution of carbon dioxide in water; or, again, by the activity of certain acids—humous acids—which originate in the decomposition of vegetable remains. These chemical agents may in some instances, _e.g._ in the case of carbonates such as limestone or dolomite—bring practically the whole rock into solution. In other instances—_e.g._ granites, basalts, etc.—they may remove some of the constituent minerals completely or partially, such as felspar, olivine, augite, and leave more resistant substances to be ultimately washed down as fine sand or mud into the river.

It is often difficult or impossible to appraise the relative efficiency of mechanical and chemical denudation in removing the materials from a certain area. There can be, indeed, little doubt that in mountainous regions the mechanical effects are largely predominant. The silts of glacial rivers are little different from freshly-powdered rock. The water which carries them but little different from the pure rain or snow which falls from the sky. There has not been time for the chemical or solvent actions to take place. Now while gravitational forces favour sudden shock and violent motions in the hills, the effect of these on solvent and chemical denudation is but small. Nor is good drainage favourable to chemical actions, for water is the primary factor in every case. Water takes up and removes soluble combinations of molecules, and penetrates beneath residual insoluble substances. It carries the oxygen and acids downwards through the soils, and finally conveys the results of its own work to the rivers and streams. The lower mean temperature of the mountains as well as the perfect drainage diminishes chemical activities.

Hence we conclude that the heights are not generally favourable to the purely solvent and chemical actions. It is on the lower-lying land that soils tend to accumulate,

and in these the chief solvent and the chief chemical denudation of the Earth are effected.

The solvent and chemical effects which go on in the finely-divided materials of the soils may be observed in the laboratory. They proceed faster than would be anticipated. The observation is made by passing a measured quantity of water backwards and forwards for some months through a tube containing a few grammes of powdered rock. Finally the water is analysed, and in this manner the amount of dissolved matter it has taken up is estimated. The rock powder is examined under the microscope in order to determine the size of the grains, and so to calculate the total surface exposed to the action of the water. We must be careful in such experiments to permit free oxidation by the atmosphere. Results obtained in this way of course take no account of the chemical effects of organic acids such as exist in the soils. The quantities obtained in the laboratory will, therefore, be deficient as compared with the natural results.

In this manner it has been found that fresh basalt exposed to continually moving water will lose about 0.20 gramme per square metre of surface per year. The mineral orthoclase, which enters largely into the constitution of many granites, was found to lose under the same conditions 0.025 gramme. A glassy lava (obsidian) rich in silica and in the chemical constituents of an average granite, was more resistant still; losing but 0.013 gramme per square metre per year. Hornblende, a mineral

abundant in many rocks, lost 0.075 gramme. The mean of the results showed that 0.08 gramme was washed in a year from each square metre. Such results give us some indication of the rate at which the work of solution goes on in the finely divided soils.[1]

It might be urged that, as the mechanical break up of rocks, and the production in this way of large surfaces, must be at the basis of solvent and chemical denudation, these latter activities should be predominant in the mountains. The answer to this is that the soils rarely owe their existence to mechanical actions. The alluvium of the valleys constitutes only narrow margins to the rivers; the finer _débris_ from the mountains is rapidly brought into the ocean. The soils which cover the greater part of continental areas have had a very different origin.

In any quarry where a section of the soil and of the underlying rock is visible, we may study the mode of formation of soils. Our observations are, we will suppose, pursued in a granite quarry. We first note that the material of the soil nearest the surface is intermixed with the roots of grasses, trees, or shrubs. Examining a handful of this soil, we see glistening flakes of mica which plainly are derived from the original granite. Washing off the finer particles, we find the largest remaining grains are composed of the all but indestructible quartz.

[1] Proc. Roy. Irish Acad., VIII., Ser. A, p. 21.

This also is from the granite. Some few of the grains are of chalky-looking felspar; again a granitic mineral. What is the finer silt we have washed off? It, too, is composed of mineral particles to a great extent; rock dust stained with iron oxide and intermixed with organic remains, both animal and vegetable. But if we make a chemical analysis of the finer silt we find that the composition is by no means that of the granite beneath. The chemist is able to say, from a study of his results, that there has been, in the first place, a large loss of material attending the conversion of the granite to the soil. He finds a concentration of certain of the more resistant substances of the granite arising from the loss of the less resistant. Thus the percentage amount of alumina is increased. The percentage of iron is also increased. But silica and most other substances show a diminished percentage. Notably lime has nearly disappeared. Soda is much reduced; so is magnesia. Potash is not so completely abstracted. Finally, owing to hydration, there is much more combined water in the soil than in the rock. This is a typical result for rocks of this kind.

Deeper in the soil we often observe a change of texture. It has become finer, and at the same time the clay is paler in colour. This subsoil represents the finer particles carried by rain from above. The change of colour is due to the state of the iron which is less oxidised low down in the soil. Beneath the subsoil the soil grows

again coarser. Finally, we recognise in it fragments of granite which ever grow larger as we descend, till the soil has become replaced by the loose and shattered rock. Beneath this the only sign of weathering apparent in the rock is the rusty hue imparted by the oxidised iron which the percolating rain has leached from iron-bearing minerals.

The soil we have examined has plainly been derived in situ from the underlying rock. It represents the more insoluble residue after water and acids have done their work. Each year there must be a very slow sinking of the surface, but the ablation is infinitesimal.

The depth of such a soil may be considerable. The total surface exposed by the countless grains of which it is composed is enormous. In a cubic foot of average soil the surface area of the grains may be 50,000 square feet or more. Hence a soil only two feet deep may expose 100,000 square feet for each square foot of surface area.

It is true that soils formed in this manner by atmospheric and organic actions take a very long time to grow. It must be remembered, however, that the process is throughout attended by the removal in solution: of chemically altered materials.

Considerations such as the foregoing must convince us that while the accumulation of the detrital sediments around the continents is largely the result of activities progressing on the steeper slopes of the land, that is,

among the mountainous regions, the feeding of the salts to the ocean arises from the slower work of meteorological and organic agencies attacking the molecular constitution of the rocks; processes which best proceed where the drainage is sluggish and the quiescent conditions permit of the development of abundant organic growth and decay.

Statistics of the solvent denudation of the continents support this view. Within recent years a very large amount of work has been expended on the chemical investigation of river waters of America and of Europe. F. W. Clarke has, at the expense of much labour, collected and compared these results. They are expressed as so many tonnes removed in solution per square mile per annum. For North America the result shows 79 tonnes so removed; for Europe 100 tonnes. Now there is a notable difference between the mean elevations of these two continents. North America has a mean elevation of 700 metres over sea level, whereas the mean elevation of Europe is but 300 metres. We see in these figures that the more mountainous land supplies less dissolved matter to the ocean than the land of lower elevation, as our study has led us to expect.

We have now considered the source of the detrital sediments, as well as of the dissolved matter which has given to the ocean, in the course of geological time, its present gigantic load of salts. It is true there are further solvent and chemical effects exerted by the sea water

upon the sediments discharged into it; but we are justified in concluding that, relatively to the similar actions taking place in the soils, the solvent and chemical work of the ocean is small. The fact is, the deposited detrital sediments around the continents occupy an area small when contrasted with the vast stretches of the land. The area of deposition is much less than that of denudation; probably hardly as much as one twentieth. And, again, the conditions of aeration and circulation which largely promote chemical and solvent denudation in the soils are relatively limited and ineffective in the detrital oceanic deposits.

The summation of the amounts of dissolved and detrital materials which denudation has brought into the ocean during the long denudative history of the Earth, as we might anticipate, reveals quantities of almost unrealisable greatness. The facts are among the most impressive which geological science has brought to light. Elsewhere in this volume they have been mentioned when discussing the age of the Earth. In the present connection, however, they are deserving of separate consideration.

The basis of our reasoning is that the ocean owes its saltness mainly if not entirely to the denudative activities we have been considering. We must establish this.

We may, in the first place, say that any other view at once raises the greatest difficulties. The chemical composition of the detrital sediments which are spread over

the continents and which build up the mountains, differs on the average very considerably from that of the igneous rocks. We know the former have been derived from the latter, and we know that the difference in the composition of the two classes of materials is due to the removal in solution of certain of the constituents of the igneous rocks. But the ocean alone can have received this dissolved matter. We know of no other place in which to look for it. It is true that some part of this dissolved matter has been again rejected by the ocean; thus the formation of limestone is largely due to the abstraction of lime from sea water by organic and other agencies. This, however, in no way relieves us of the necessity of tracing to the ocean the substances dissolved from the igneous rocks. It follows that we have here a very causa for the saltness of the ocean. The view that the ocean "was salt from the first" is without one known fact to support it, and leaves us with the burden of the entire dissolved salts of geological time to dispose of—Where and how?

The argument we have outlined above becomes convincingly strong when examined more closely. For this purpose we first compare the average chemical composition of the sedimentary and the igneous rocks. The following table gives the percentages of the chief chemical constituents: [1]

[1] F. W. Clarke: _A Preliminary Study of Chemical Denudation_, p. 13

	Igneous.	Sedimentary.
Silica (SiO_2) -	59.99	58.51
Alumina (Al_2O_3) -	15.04	13.07
Ferric oxide (F_2O_3) -	2.59	3.40
Ferrous oxide (FeO) -	3.34	2.00

Magnesia (MgO) -	3.89	2.52
Lime (CaO) -	4.81	5.42
Soda (Na_2O) -	3.41	1.12
Potash (K_2O) -	2.95	2.80
Water (H_2O) -	1.92	4.28
Carbon dioxide (CO_2) -	--	4.93
Minor constituents -	2.06	1.95
	100.00	100.00

In the derivation of the sediments from the igneous rocks there is a loss by solution of about 33 per cent; _i.e._ 100 tons of igneous rock yields rather less than 70 tons of sedimentary rock. This involves a concentration in the sediments of the more insoluble constituents. To this rule the lime-content appears to be an exception. It is not so in reality. Its high value in the sediments is due to its restoration from the ocean to the land. The magnesia and potash are, also, largely restored from the ocean; the former in dolomites and magnesian limestones; the latter in glauconite sands. The iron of the sediments shows increased oxidation. The most notable difference in the two analyses appears, however, in the soda percentages. This falls from 3.41 in the igneous rock to 1.12 in the average sediment. Indeed, this

deficiency of soda in sedimentary rocks is so characteristic of secondary rocks that it may with some safety be applied to discriminate between the two classes of substances in cases where petrological distinctions of other kinds break down.

To what is this so marked deficiency of soda to be ascribed? It is a result of the extreme solubility of the salts of sodium in water. This has not only rendered its deposition by evaporation a relatively rare and unimportant incident of geological history, but also has protected it from abstraction from the ocean by organic agencies. The element sodium has, in fact, accumulated in the ocean during the whole of geological time.

We can use the facts associated with the accumulation of sodium salts in the ocean as a means of obtaining additional support to the view, that the processes of solvent denudation are responsible for the saltness of the ocean. The new evidence may be stated as follows: Estimates of the amounts of sedimentary rock on the continents have repeatedly been made. It is true that these estimates are no more than approximations. But they undoubtedly _are_ approximations, and as such may legitimately be used in our argument; more especially as final agreement tends to check and to support the several estimates which enter into them.

The most recent and probable estimates of the sediments on the land assign an average thickness of one mile of

secondary rocks over the land area of the world. To this some increase must be made to allow for similar materials concealed in the ocean, principally around the continental margins. If we add 10 per cent. and assign a specific gravity of 2.5 we get as the mass of the sediments 64×10^{16} tonnes. But as this is about 67 per cent. of the parent igneous rock—_i.e._ the average igneous rock from which the sediments are derived—we conclude that the primary denuded rock amounted to a mass of about 95×10^{16} tonnes.

Now from the mean chemical composition of the secondary rocks we calculate that the mass of sediments as above determined contains 0.72×10^{16} tonnes of the sodium oxide, Na_2O. If to this amount we add the quantity of sodium oxide which must have been given to the ocean in order to account for the sodium salts contained therein, we arrive at a total quantity of oxide of sodium which must be that possessed by the primary rock before denudation began its work upon it. The mass of the ocean being well ascertained, we easily calculate that the sodium in the ocean converted to sodium oxide amounts to 2.1×10^{16} tonnes. Hence between the estimated sediments and the waters of the ocean we can account for 2.82×10^{16} tonnes of soda. When now we put this quantity back into the estimated mass of primary rock we find that it assigns to the primary rock a soda percentage of 3.0. On the average analysis given above this should be 3.41 per cent. The agreement,

all things considered, more especially the uncertainty in the estimate of the sediments, is plainly in support of the view that oceanic salts are derived from the rocks; if, indeed, it does not render it a certainty.

A leading and fundamental inference in the denudative history of the Earth thus finds support: indeed, we may say, verification. In the light of this fact the whole work of denudation stands revealed. That the ocean began its history as a vast fresh-water envelope of the Globe is a view which accords with the evidence for the primitive high temperature of the Earth. Geological history opened with the condensation of an atmosphere of immense extent, which, after long fluctuations between the states of steam and water, finally settled upon the surface, almost free of matter in solution: an ocean of distilled water. The epoch of denudation then began. It will, probably, continue till the waters, undergoing further loss of thermal energy, suffer yet another change of state, when their circulation will cease and their attack upon the rocks come to an end.

From what has been reviewed above it is evident that the sodium in the ocean is an index of the total activity of denudation integrated over geological time. From this the broad facts of the results of denudation admit of determination with considerable accuracy. We can estimate the amount of rock which has been degraded by solvent and chemical actions, and the amount of sediments which has been derived from it. We are,

thus, able to amend our estimate of the sediments which, as determined by direct observation, served to support the basis of our argument.

We now go straight to the ocean for the amount of sodium of denudative origin. There may, indeed, have been some primitive sodium dissolved by a more rapid denudation while the Earth's surface was still falling in temperature. It can be shown, however, that this amount was relatively small. Neglecting it we may say with safety that the quantity of sodium carried into the ocean by the rivers must be between 14,000 and 15,000 million million tonnes: _i.e._ 14,500 x 10^{12} tonnes, say.

Keeping the figures to round numbers we find that this amount of sodium involves the denudation of about 80 x 10^{16} tonnes of average igneous rock to 53 x 1016 tonnes of average sediment. From these vast quantities we know that the parent rock denuded during geological time amounted to some 300 million cubic kilometres or about seventy million cubic miles. The sediments derived therefrom possessed a bulk of 220 million cubic kilometres or fifty million cubic miles. The area of the land surface of the Globe is 144 million square kilometres. The parent rock would have covered this to a uniform depth of rather more than two kilometres, and the derived sediment to more than 1.5 kilometres, or about one mile deep.

The slow accomplishment of results so vast conveys some idea of the great duration of geological time.

47

The foregoing method of investigating the statistics of solvent denudation is capable of affording information not only as to the amount of sediments upon the land, but also as to the quantity which is spread over the floor of the ocean.

We see this when we follow the fate of the 33 per cent. of dissolved salts which has been leached from the parent igneous rock, and the mass of which we calculate from the ascertained mass of the latter, to be 27×10^{16} tonnes. This quantity was at one time or another all in the ocean. But, as we saw above, a certain part of it has been again abstracted from solution, chiefly by organic agencies. Now the abstracted solids have not been altogether retained beneath the ocean. Movements of the land during geological time have resulted in some portion being uplifted along with other sediments. These substances constitute, mainly, the limestones.

We see, then, that the 27×10^{16} tonnes of substances leached from the parent igneous rocks have had a threefold destination. One part is still in solution; a second part has been precipitated to the bottom of the ocean; a third part exists on the land in the form of calcareous rocks.

Observation on the land sediments shows that the calcareous rocks amount to about 5 per cent. of the whole. From this we find that 3×10^{16} tonnes, approximately, of such rocks have been taken from the ocean. This accounts for one of the three classes of material

into which the original dissolved matter has been divided. Another of the three quantities is easily estimated: the amount of matter still in solution in the ocean. The volume of the ocean is 1,414 million cubic kilometres and its mass is 145×10^{16} tonnes. The dissolved salts in it constitute 3.4 per cent. of its mass; or, rather more than 5×10^{16} tonnes. The limestones on the land and the salts in the sea water together make up about 8×10^{16} tonnes. If we, now, deduct this from the total of 27×10^{16} tonnes, we find that about 19×10^{16} tonnes must exist as precipitated matter on the floor of the ocean.

The area of the ocean is 367×10^{12} square metres, so that if the precipitated sediment possesses an average specific gravity of 2.5, it would cover the entire floor to a uniform depth of 218 metres; that is 715 feet. This assumes that there was uniform deposition of the abstracted matter over the floor of the ocean. Of course, this assumption is not justifiable. It is certain that the rate of deposition on the floor of the sea has varied enormously with various conditions—principally with the depth. Again, it must be remembered that this estimate takes no account of solid materials otherwise brought into the oceanic deposits; _e.g._, by wind-transported dust from the land or volcanic ejectamenta in the ocean depths. It is not probable, however, that any considerable addition to the estimated mean depth of deposit from such sources would be allowable.

The greatness of the quantities involved in these determinations is almost awe inspiring. Take the case of the dissolved salts in the ocean. They are but a fraction, as we have seen, of the total results of solvent denudation and represent the integration of the minute traces contributed by the river water. Yet the common salt (chloride of sodium) alone, contained in the ocean, would, if abstracted and spread over the dry land as a layer of rock salt having a specific gravity of 2.2, cover the whole to a depth of 107 metres or 354 feet. The total salts in solution in the ocean similarly spread over the land would increase the depth of the layer to 460 feet. After considering what this means we have to remember that this amount of matter now in solution in the seas is, in point of fact, less than a fifth part of the total dissolved from the rocks during geological time.

The transport by denudation of detrital and dissolved matter from the land to the ocean has had a most important influence on the events of geological history. The existing surface features of the earth must have been largely conditioned by the dynamical effects arising therefrom. In dealing with the subject of mountain genesis we will, elsewhere, see that all the great mountain ranges have originated in the accumulation of the detrital sediments near the shore in areas which, in consequence of the load, gradually became depressed and developed into synclines of many thousands of feet in depth. The most impressive surface features of the Globe originated

in this manner. We will see too that these events were of a rhythmic character; the upraising of the mountains involving intensified mechanical denudation over the elevated area and in this way an accelerated transport of detritus to the sea; the formation of fresh deposits; renewed synclinal sinking of the sea floor, and, finally, the upheaval of a younger mountain range. This extraordinary sequence of events has been determined by the events of detrital denudation acting along with certain general conditions which have all along involved the growth of compressive stresses in the surface crust of the Earth.

The effects of purely solvent denudation are less easily traced, but, very probably, they have been of not less importance. I refer here to the transport from the land to the sea of matter in solution.

Solvent denudation, as observed above, takes place mainly in the soils and in this way over the more level continental areas. It has resulted in the removal from the land and transfer to the ocean of an amount of matter which represents a uniform layer of one half a kilometre; that is of more than 1,600 feet of rock. The continents have, during geological time, been lightened to this extent. On the other hand all this matter has for the greater part escaped the geosynclines and become uniformly diffused throughout the ocean or precipitated over its floor principally on the continental slopes before the great depths are reached. Of this material the ocean

waters contain in solution an amount sufficient to increase their specific gravity by 2.7 per cent.

Taking the last point first, it is interesting to note the effects upon the bulk of the ocean which has resulted from the matter dissolved in it. From the known density of average sea water we find that 100 ccs. of it weigh just 102.7 grammes. Of this 3.5 per cent. by weight are solids in solution. That is to say, 3.594 grammes. Hence the weight of water present is 99.1 grammes, or a volume of 99.1 ccs. From this we see that the salts present have increased the volume by 0.9 ccs. or 0.9 per cent.

The average depth of the ocean is 2,000 fathoms or 3,700 metres. The increase of depth due to salts dissolved in the ocean has been, therefore, 108 feet or 33.24 metres. This result assumes that there has been no increased elastic compression due to the increased pressure, and no change of compressional elastic properties. We may be sure that the rise on the shore line of the land has not been less than 100 feet.

We see then that as the result of solvent denudation we have to do with a heavier and a deeper ocean, expanded in volume by nearly one per cent. and the floor of which has become raised, on an average, about 700 feet by precipitated sediment.

One of the first conceptions, which the student of geology has to dismiss from his mind, is that of the immobility or rigidity of the Earth's crust. The lane, we live on sways even to the gentle rise and fail of ocean tides

around the coasts. It suffers its own tidal oscillations due to the moon's attractions. Large tracts of semi-liquid matter underlie it. There is every evidence that the raised features of the Globe are sustained by such pressures acting over other and adjacent areas as serve to keep them in equilibrium against the force of gravity. This state of equilibrium, which was first recognised by Pratt, as part of the dynamics of the Earth's crust, has been named isostasy. The state of the crust is that of "mobile equilibrium."

The transfer of matter from the exposed land surfaces to the sub-oceanic slopes of the continents and the increase in the density of the ocean, must all along have been attended by isostatic readjustment. We cannot take any other view. On the one hand the land was being lightened; on the other the sea was increasing in mass and depth and the flanks of the continents were being loaded with the matter removed from the land and borne in solution to the ocean. How important the resulting movements must have been may be gathered from the fact that the existing land of the Globe stands at a mean elevation of no more than 2,000 feet above sea level. We have seen that solvent denudation removed over 1,600 feet of rock. But we have no evidence that on the whole the elevation of land in the past was ever very different from what it now is.

We have, then, presented to our view the remarkable fact that throughout the past, and acting with extreme

slowness, the land has steadily been melted down into the sea and as steadily been upraised from the waters. It is possible that the increased bulk of the ocean has led to a certain diminution of the exposed land area. The point is a difficult one. One thing we may without much risk assume. The sub-aereal current of dissolved matter from the land to the ocean was accompanied by a sub-crustal flux from the ocean areas to the land areas; the heated viscous materials creeping from depths far beneath the ocean floor to depths beneath the roots of the mountains which arose around the oceans. Such movements took ages for their accomplishment. Indeed, they have been, probably, continuous all along and are still proceeding. A low degree of viscosity will suffice to permit of movements so slow. Superimposed upon these movements the rhythmic alternations of depression and elevation of the geosynclines probably resulted in releasing the crust from local accumulation of strains arising in the more rigid surface materials. The whole sequence of movements presents an extraordinary picture of pseudo-vitality—reminding us of the circulatory and respiratory systems of a vast organism.

All great results in our universe are founded in motions and forces the most minute. In contemplating the Cause or the Effect we stand equally impressed with the spectacle presented to us. We shall now turn from the great effects of denudation upon the history and evolution of a world and consider for a moment activities

so minute in detail that their operations will probably for ever elude our bodily senses, but which nevertheless have necessarily affected and modified the great results we have been considering.

The ocean a little way from the land is generally so free from suspended sediments that it has a blackness as of ink. This blackness is due to its absolute freedom from particles reflecting the sun's light. The beautiful blue of the Swiss and Italian lakes is due to the presence of very fine particles carried into them by the rivers; the finest flour of the glaciers, which remain almost indefinitely suspended in the water. But in the ocean it is only in those places where rapid currents running over shallows stir continually the sediments or where the fresh water of a great river is carried far from the land, that the presence of silt is to be observed. The beautiful phenomenon of the coal-black sea is familiar to every yachtsman who has sailed to the west of our Islands.[1]

There is, in fact, a very remarkable difference in the manner of settlement of fine sediments in salt and in fresh water. We are here brought into contact with one of those subtle yet influential natural actions the explanation of which involves scientific advance along many apparently unconnected lines of investigation.

[1] See Tyndall's Voyage to Algeria in _Fragments of Science._ The cause of the blue colour of the lakes has been discussed by various observers, not always with agreement.

It is easy to observe in the laboratory the fact of the different behaviour of salt and fresh water towards finely divided substances. The nature of the insoluble substance is not important.

We place, in a good light, two glass vessels of equal dimensions; the one filled with sea water, the other with fresh water. Into each we stir the same weight of very finely powdered slate: just so much as will produce a cloudiness. In a few hours we find the sea water limpid. The fresh water is still cloudy, however; and, indeed, may be hardly different in appearance from what it was at starting. In itself this is a most extraordinary experiment. We would have anticipated quite the opposite result owing to the greater density of the sea water.

But a still more interesting experiment remains to be carried out. In the sea water we have many different salts in solution. Let us see if these salts are equally responsible for the result we have obtained. For this purpose we measure out quantities of sodium chloride and magnesium chloride in the proportion in which they exist in sea water: that is about as seven to one. We add such an equal amount of water to each as represents the dilution of these salts in sea water. Then finally we stir a little of the finely powdered slate into each. It will be found that the magnesium chloride, although so much more dilute than the sodium chloride, is considerably more active in clearing out the suspension. We may now try such marine salts as magnesium sulphate,

or calcium sulphate against sodium chloride; keeping the marine proportions. Again we find that the magnesium and calcium salts are the most effective, although so much more dilute than the sodium salt.

There is no visible clue to the explanation of these results. But we must conclude as most probable that some action is at work in the sea water and in the salt solutions which clumps or flocculates the sediment. For only by the gathering of the particles together in little aggregates can we explain their rapid fall to the bottom. It is not a question of viscosity (_i.e._ of resistance to the motion of the particles), for the salt solutions are rather more viscous than the fresh water. Still more remarkable is the fact that every dissolved substance will not bring about the result. Thus if we dissolve sugar in water we find that, if anything, the silt settles more slowly in the sugar solution than in fresh water.

Now there is one effect produced by the solution of such salts as we have dealt with which is not produced by such bodies as sugar. The water is rendered a conductor of electricity. Long ago Faraday explained this as due to the presence of free atoms of the dissolved salt in the solution, carrying electric charges. We now speak of the salt as "ionised." That is it is partly split up into ions or free electrified atoms of chlorine, sodium, magnesium, etc., according to the particular salt in solution. This fact leads us to think that these electrified

atoms moving about in the solution may be the cause of the clumping or flocculation. Such electrified atoms are absent from the sugar solution: sugar does not become "ionised" when it is dissolved.

The suspicion that the free electrified atoms play a part in the phenomenon is strengthened when we recall the remarkable difference in the action of sodium chloride and magnesium chloride. In each of the solutions of these substances there are free chlorine atoms each of which carries a single charge of negative electricity. As these atoms are alike in both solutions the different behaviour of the solutions cannot be due to the chlorine. But the metallic atom is very different in the two cases. The ionised sodium atom is known to be _monad_ or carries but _one_ positive charge; whereas the magnesium atom is _diad_ and carries _two_ positive charges. If, then, we assume that the metallic, positively electrified atom is in each case responsible, we have something to go on. It may be now stated that it has been found by experiment and supported by theory that the clumping power of an ion rises very rapidly with its valency; that is with the number of unit charges associated with it. Thus diads such as magnesium, calcium, barium, etc., are very much more efficient than monads such as sodium, potassium, etc., and again, triads such as aluminium are, similarly, very much more powerful than diad atoms. Here, in short, we have arrived at the active cause of the phenomenon. Its inner mechanism

is, however, harder to fathom. A plausible explanation can be offered, but a study of it would take us too far. Sufficient has been said to show the very subtile nature of the forces at work.

We have here an effect due to the sea salts derived by denudation from the land which has been slowly augmenting during geological time. It is certain that the ocean was practically fresh water in remote ages. During those times the silt from the great rivers would have been carried very far from the land. A Mississippi of those ages would have sent its finer suspensions far abroad on a contemporary Gulf stream: not improbably right across the Atlantic. The earlier sediments of argillaceous type were not collected in the geosynclines and the genesis of the mountains was delayed proportionately. But it was, probably, not for very long that such conditions prevailed. For the accumulation of calcium salts must have been rapid, and although the great salinity due to sodium salts was of slow growth the salts of the diad element calcium must have soon introduced the cooperation of the ion in the work of building the mountain.

THE ABUNDANCE OF LIFE [1]

WE had reached the Pass of Tre Croci[2]and from a point a little below the summit, looked eastward over the glorious Val Buona. The pines which clothed the floor and lower slopes of the valley, extended their multitudes into the furthest distance, among the many recesses of the mountains, and into the confluent Val di Misurina. In the sunshine the Alpine butterflies flitted from stone to stone. The ground at our feet and everywhere throughout the forests teamed with the countless millions of the small black ants.

It was a magnificent display of vitality; of the aggressiveness of vitality, assailing the barren heights of the limestone, wringing a subsistence from dead things. And the question suggested itself with new force: why the abundance of life and its unending activity?

In trying to answer this question, the present sketch originated.

I propose to refer for an answer to dynamic considerations. It is apparent that natural selection can only be concerned in a secondary way. Natural selection defines

[1] Proc. Roy. Dublin Soc., vol. vii., 1890.

[2] In the Dolomites of Southeast Tyrol; during the summer of 1890. Much of what follows was evolved in discussion with my fellow-traveller, Henry H. Dixon. Much of it is his.

a certain course of development for the organism; but very evidently some property of inherent progressiveness in the organism must be involved. The mineral is not affected by natural selection to enter on a course of continual variation and multiplication. The dynamic relations of the organism with the environment are evidently very different from those of inanimate nature.

GENERAL DYNAMIC CONDITIONS ATTENDING INANIMATE ACTIONS

It is necessary, in the first place, to refer briefly to the phenomena attending the transfer of energy within and into inanimate material systems. It is not assumed here that these phenomena are restricted in their sphere of action to inanimate nature. It is, in fact, very certain that they are not; but while they confer on dead nature its own dynamic tendencies, it will appear that their effects are by various means evaded in living nature. We, therefore, treat of them as characteristic of inanimate actions. We accept as fundamental to all the considerations which follow the truth of the principle of the Conservation of Energy.[1]

[1] "The principle of the Conservation of Energy has acquired so much scientific weight during the last twenty years that no physiologist would feel any confidence in an experiment which showed a considerable difference between the work done by the animal and the balance of the account of Energy received and spent."—Clerk Maxwell, _Nature_, vol. xix., p. 142. See also Helmholtz _On the Conservation of Force._

Whatever speculations may be made as to the course of events very distant from us in space, it appears certain that dissipation of energy is at present actively progressing throughout our sphere of observation in inanimate nature. It follows, in fact, from the second law of thermodynamics, that whenever work is derived from heat, a certain quantity of heat falls in potential without doing work or, in short, is dissipated. On the other hand, work may be entirely converted into heat. The result is the heat-tendency of the universe. Heat, being an undirected form of energy, seeks, as it were, its own level, so that the result of this heat-tendency is continual approach to uniformity of potential.

The heat-tendency of the universe is also revealed in the far-reaching "law of maximum work," which defines that chemical change, accomplished without the intervention of external energy, tends to the production of the body, or system of bodies, which disengage the greatest quantity of heat.[1] And, again, vast numbers of actions going on throughout nature are attended by dissipatory thermal effects, as those arising from the motions of proximate molecules (friction, viscosity), and from the fall of electrical potential.

Thus, on all sides, the energy which was once most probably existent in the form of gravitational potential, is being dissipated into unavailable forms. We must

[1] Berthelot, _Essai de Mécanique Chimique._

recognize dissipation as an inevitable attendant on inanimate transfer of energy.

But when we come to consider inanimate actions in relation to time, or time-rate of change, we find a new feature in the phenomena attending transfer of energy; a feature which is really involved in general statements as to the laws of physical interactions.[1] It is seen, that the attitude of inanimate material systems is very generally, if not in all cases, retardative of change—opposing it by effects generated by the primary action, which may be called "secondary" for convenience. Further, it will be seen that these secondary effects are those concerned in bringing about the inevitable dissipation.

As example, let us endeavour to transfer gravitational potential energy contained in a mass raised above the surface of the Earth into an elastic body, which we can put into compression by resting the weight upon it. In this way work is done against elastic force and stored as elastic potential energy. We may deal with a metal spring, or with a mass of gas contained in a cylinder fitted with a piston upon which the weight may be placed. In either case we find the effect of compression is to raise the temperature of the substance, thus causing its

[1] Helmholtz, _Ice and Glaciers._ Atkinson's collection of his Popular Lectures. First Series, p.120. Quoted by Tate, _Heat_, p. 311.

expansion or increased resistance to the descent of the weight. And this resistance continues, with diminishing intensity, till all the heat generated is dissipated into the surrounding medium. The secondary effect thus delays the final transfer of energy.

Again, if we suppose the gas in the cylinder replaced by a vapour in a state of saturation, the effect of increased pressure, as of a weight placed upon the piston, is to reduce the vapour to a liquid, thereby bringing about a great diminution of volume and proportional loss of gravitational potential by the weight. But this change will by no means be brought about instantaneously. When a little of the vapour is condensed, this portion parts with latent heat of vaporisation, increasing the tension of the remainder, or raising its point of saturation, so that before the weight descends any further, this heat has to escape from the cylinder.

Many more such cases might be cited. The heating of india-rubber when expanded, its cooling when compressed, is a remarkable one; for at first sight it appears as if this must render it exceptional to the general law, most substances exhibiting the opposite thermal effects when stressed. However, here, too, the action of the stress is opposed by the secondary effects developed in the substance; for it is found that this substance contracts when heated, expands when cooled. Again, ice being a substance which contracts in melting, the effect of pressure is to facilitate melting, lowering its freezing point. But

so soon as a little melting occurs, the resulting liquid calls on the residual ice for an amount of heat equivalent to the latent heat of liquefaction, and so by cooling the whole, retards the change.

Such particular cases illustrate a principle controlling the interaction of matter and energy which seems universal in application save when evaded, as we shall see, by the ingenuity of life. This principle is not only revealed in the researches of the laboratory; it is manifest in the history of worlds and solar systems. Thus, consider the effects arising from the aggregation of matter in space under the influence of the mutual attraction of the particles. The tendency here is loss of gravitational potential. The final approach is however retarded by the temperature, or vis viva of the parts attending collision and compression. From this cause the great suns of space radiate for ages before the final loss of potential is attained.

Clerk Maxwell[1] observes on the general principle that less force is required to produce a change in a body when the change is unopposed by constraints than when it is subjected to such. From this if we assume the external forces acting upon a system not to rise above a certain potential (which is the order of nature), the constraints of secondary actions may, under certain circumstances, lead to final rejection of some of the energy, or, in any

[1] _Theory of Heat_, p. 131.

case, to retardation of change in the system—dissipation of energy being the result.[1]

As such constraints seem inherently present in the properties of matter, we may summarise as follows:

The transfer of energy into any inanimate material system is attended by effects retardative to the transfer and conducive to dissipation.

Was this the only possible dynamic order ruling in material systems it is quite certain the myriads of ants and pines never could have been, except all generated by creative act at vast primary expenditure of energy. Growth and reproduction would have been impossible in systems which retarded change at every step and never proceeded in any direction but in that of dissipation. Once created, indeed, it is conceivable that, as heat engines, they might have dragged out an existence of alternate life and death; life in the hours of sunshine, death in hours of darkness: no final death, however, their lot, till their parts were simply worn out by long use, never made good by repair. But the sustained and increasing activity of organized nature is a fact; therefore some other order of events must be possible.

[1] The law of Least Action, which has been applied, not alone in optics, but in many mechanical systems, appears physically based upon the restraint and retardation opposing the transfer of energy in material systems.

GENERAL DYNAMIC CONDITIONS ATTENDING ANIMATE ACTIONS

What is the actual dynamic attitude of the primary organic engine—the vegetable organism? We consider, here, in the first place, not intervening, but resulting phenomena.

The young leaf exposed to solar radiation is small at first, and the quantity of radiant energy it receives in unit of time cannot exceed that which falls upon its surface. But what is the effect of this energy? Not to produce a retardative reaction, but an accelerative response: for, in the enlarging of the leaf by growth, the plant opens for itself new channels of supply.

If we refer to "the living protoplasm which, with its unknown molecular arrangement, is the only absolute test of the cell and of the organism in general,[1] we find a similar attitude towards external sources of available energy. In the act of growth increased rate of assimilation is involved, so that there is an acceleration of change till a bulk of maximum activity is attained. The surface, finally, becomes too small for the absorption of energy adequate to sustain further increase of mass (Spencer[2]), and the acceleration ceases. The waste going on in the central parts is then just balanced by the renewal at the surface. By division, by spreading of the mass, by

[1] Claus, _Zoology_, p. 13.

[2] Geddes and Thomson, _The Evolution of Sex_, p. 220.

out-flowing processes, the normal activity of growth may be restored. Till this moment nothing would be gained by any of these changes. One or other of them is now conducive to progressive absorption of energy by the organism, and one or other occurs, most generally the best of them, subdivision. Two units now exist; the total mass immediately on division is unaltered, but paths for the more abundant absorption of energy are laid open.

The encystment of the protoplasm (occurring under conditions upon which naturalists do not seem agreed[1]) is to all appearance protective from an unfavourable environment, but it is often a period of internal change as well, resulting in a segregation within the mass of numerous small units, followed by a breakup of the whole into these units. It is thus an extension of the basis of supply, and in an impoverished medium, where unit of surface is less active, is evidently the best means of preserving a condition of progress.

Thus, in the organism which forms the basis of all modes of life, a definite law of action is obeyed under various circumstances of reaction with the available energy of its environment.

Similarly, in the case of the more complex leaf, we see, not only in the phenomenon of growth, but in its extension in a flattened form, and in the orientation of greatest surface towards the source of energy, an attitude towards

[1] However, "In no way comparable with death." Weismann, _Biological Memoirs_, p. 158.

available energy causative of accelerated transfer. There is seemingly a principle at work, leading to the increase of organic activity.

Many other examples might be adduced. The gastrula stage in the development of embryos, where by invagination such an arrangement of the multiplying cells is secured as to offer the greatest possible surface consistent with a first division of labour; the provision of cilia for drawing upon the energy supplies of the medium; and more generally the specialisation of organs in the higher developments of life, may alike be regarded as efforts of the organism directed to the absorption of energy. When any particular organ becomes unavailing in the obtainment of supplies, the organ in the course of time becomes aborted or disappears.[1] On the other hand, when a too ready and liberal supply renders exertion and specialisation unnecessary, a similar abortion of functionless organs takes place. This is seen in the degraded members of certain parasites.

During certain epochs of geological history, the vegetable world developed enormously; in response probably to liberal supplies of carbon dioxide. A structural adaptation to the rich atmosphere occurred, such as was calculated to cooperate in rapidly consuming the supplies, and to this obedience to a law of progressive transfer of energy we owe the vast stores of energy now accumulated

[1] Claus, _Zoology_, p. 157

in our coal fields. And when, further, we reflect that this store of energy had long since been dissipated into space but for the intervention of the organism, we see definitely another factor in organic transfer of energy—a factor acting conservatively of energy, or antagonistically to dissipation.

The tendency of organized nature in the presence of unlimited supplies is to "run riot." This seems so universal a relation, that we are safe in seeing here cause and effect, and in drawing our conclusions as to the attitude of the organism towards available energy. New species, when they come on the field of geological history, armed with fresh adaptations, irresistible till the slow defences of the subjected organisms are completed, attain enormous sizes under the stimulus of abundant supply, till finally, the environment, living and dead, reacts upon them with restraining influence. The exuberance of the organism in presence of energy is often so abundant as to lead by deprivation to its self-destruction. Thus the growth of bacteria is often controlled by their own waste products. A moment's consideration shows that such progressive activity denotes an accelerative attitude on the part of the organism towards the transfer of energy into the organic material system. Finally, we are conscious in ourselves how, by use, our faculties are developed; and it is apparent that all such progressive developments must rest on actions which respond to supplies with fresh demands. Possibly in the present and ever-increasing consumption of inanimate power by civilised races, we see revealed the dynamic attitude of the organism working through thought-processes.

Whether this be so or not, we find generally in organised nature causes at work which in some way lead to a progressive transfer of energy into the organic system. And we notice, too, that all is not spent, but both immediately in the growth of the individual, and ultimately in the multiplication of the species, there are actions associated with vitality which retard the dissipation of energy. We proceed to state the dynamical principles involved in these manifestations, which appear characteristic of the organism, as follows:—

The transfer of energy into any animate material system is attended by effects conducive to the transfer, and retardative of dissipation.

This statement is, I think, perfectly general. It has been in part advanced before, but from the organic more than the physical point of view. Thus, "hunger is an essential characteristic of living matter"; and again, "hunger is a dominant characteristic of living matter,"[1] are, in part, expressions of the statement. If it be objected against the generality of the statement, that there are periods in the life of individuals when stagnation and decay make their appearance, we may answer, that

[1] _Evolution of Sex._ Geddes and Thomson, chap. xvi. See also a reference to Cope's theory of "Growth Force," in Wallace's _Darwinism_, p. 425.

such phenomena arise in phases of life developed under conditions of external constraint, as will be urged more fully further on, and that in fact the special conditions of old age do not and cannot express the true law and tendency of the dynamic relations of life in the face of its evident advance upon the Earth. The law of the unconstrained cell is growth on an ever increasing scale; and although we assume the organic configuration, whether somatic or reproductive, to be essentially unstable, so that continual inflow of energy is required merely to keep it in existence, this does not vitiate the fact that, when free of all external constraint, growth gains on waste. Indeed, even in the case of old age, the statement remains essentially true, for the phenomena then displayed point to a breakdown of the functioning power of the cell, an approximation to configurations incapable of assimilation. It is not as if life showed in these phenomena that its conditions could obtain in the midst of abundance, and yet its law be suspended; but as if they represented a degradation of the very conditions of life, a break up, under the laws of the inanimate, of the animate contrivance; so that energy is no longer available to it, or the primary condition, "the transfer of energy into the animate system," is imperfectly obeyed. It is to the perfect contrivance of life our statement refers.

That the final end of all will be general non-availability there seems little reason to doubt, and the organism, itself dependent upon differences of potential, cannot

hope to carry on aggregation of energy beyond the period when differences of potential are not. The organism is not accountable for this. It is being affected by events external to it, by the actions going on through inanimate agents. And although there be only a part of the received energy preserved, there is a part preserved, and this amount is continually on the increase. To see this it is only necessary to reflect that the sum of animate energy—capability of doing work in any way through animate means—at present upon the Earth, is the result, although a small one, of energy reaching the Earth since a remote period, and which otherwise had been dissipated in space. In inanimate actions throughout nature, as we know it, the availability is continually diminishing. The change is all the one way. As, however, the supply of available energy in the universe is (probably) limited in amount, we must look upon the two as simply effecting the final dissipation of potential in very different ways. The animate system is aggressive on the energy available to it, spends with economy, and invests at interest till death finally deprives it of all. It has heirs, indeed, who inherit some of its gains, but they, too, must die, and ultimately there will be no successors, and the greater part must melt away as if it had never been. The inanimate system responds to the forces imposed upon it by sluggish changes; of that which is thrust upon it, it squanders uselessly. The path of the energy is very different in the two cases.

While it is true generally that both systems ultimately result in the dissipation of energy to uniform potential, the organism can, as we have seen, under particular circumstances evade the final doom altogether. It can lay up a store of potential energy which may be permanent. Thus, so long as there is free oxygen in the universe, our coalfields might, at any time in the remote future, generate light and heat in the universal grave.

It is necessary to observe on the fundamental distinction between the growth of the protoplasm and the growth of the crystal. It is common to draw comparison between the two, and to point to metabolism as the chief distinction. But while this is the most obvious distinction the more fundamental one remains in the energy relations of the two with the environment.[1] The growth of the crystal is the result of loss of energy; that of the organism the result of gain of energy. The crystal represents a last position of stable equilibrium assumed by molecules upon a certain loss of kinetic energy, and the formation of the crystal by evaporation and concentration of a liquid does not, in its dynamic aspect, differ much from the precipitation of an amorphous sediment. The organism, on the other hand, represents a more or less unstable condition formed and maintained by inflow of energy; its formation, indeed, often attended with a loss of kinetic energy (fixation of carbon in plants), but, if so, accompanied by

[1] It appears exceptional for the crystal line configuration to stand higher in the scale of energy than the amorphous.

a more than compensatory increase of potential molecular energy.

Thus, between growth in the living world and growth in the dead world, the energy relations with the environment reveal a marked contrast. Again, in the phenomena of combustion, there are certain superficial resemblances which have led to comparison between the two. Here again, however, the attitudes towards the energy of the environment stand very much as + and -. The life absorbs, stores, and spends with economy. The flame only recklessly spends. The property of storage by the organism calls out a further distinction between the course of the two processes. It secures that the chemical activity of the organism can be propagated in a medium in which the supply of energy is discontinuous or localised. The chemical activity of the combustion can, strictly speaking, only be propagated among contiguous particles. I need not dwell on the latter fact; an example of the former is seen in the action of the roots of plants, which will often traverse a barren place or circumvent an obstacle in their search for energy. In this manner roots will find out spots of rich nutriment.

Thus there is a dynamic distinction between the progress of the organism and the progress of the combustion, or of the chemical reaction generally. And although there be unstable chemical systems which absorb energy during reaction, these are (dynamically) no more than the expansion of the compressed gas. There is a certain

initial capacity in the system for a given quantity of energy; this satisfied, progress ceases. The progress of the organism in time is continual, and goes on from less to greater so long as its development is unconstrained and the supply of energy is unlimited.

We must regard the organism as a configuration which is so contrived as to evade the tendency of the universal laws of nature. Except we are prepared to believe that a violation of the second law of thermodynamics occurs in the organism, that a "sorting demon" is at work within it, we must, I think, assume that the interactions going on among its molecules are accompanied by retardation and dissipation like the rest of nature. That such conditions are not incompatible with the definition of the dynamic attitude of the organism, can be shown by analogy with our inanimate machines which, by aid of hypotheses in keeping with the second law of thermodynamics, may be supposed to fulfil the energy-functions of the plant or animal, and, in fact, in all apparent respects conform to the definition of the organism.

We may assume this accomplished by a contrivance of the nature of a steam-engine, driven by solar energy. It has a boiler, which we may suppose fed by the action of the engine. It has piston, cranks, and other movable parts, all subject to resistance from friction, etc. Now there is no reason why this engine should not expend its surplus energy in shaping, fitting, and starting into action other engines:—in fact, in reproductive sacrifice. All

these other engines represent a multiplied absorption of energy as the effects of the energy received by the parent engine, and may in time be supposed to reproduce themselves. Further, we may suppose the parent engine to be small and capable of developing very little power, but the whole series as increasing in power at each generation. Thus the primary energy relations of the vegetable organism are represented in these engines, and no violation of the second law of thermodynamics involved.

We might extend the analogy, and assuming these engines to spend a portion of their surplus energy in doing work against chemical forces—as, for example, by decomposing water through the intervention of a dynamo—suppose them to lay up in this way a store of potential energy capable of heating the boilers of a second order of engines, representing the graminivorous animal. It is obvious without proceeding to a tertiary or carnivorous order, that the condition of energy in the animal world may be supposed fulfilled in these successive series of engines, and no violation of the principles governing the actions going on in our machines assumed. Organisms evolving on similar principles would experience loss at every transfer. Thus only a portion of the radiant energy absorbed by the leaf would be expended in actual work, chemical and gravitational, etc. It is very certain that this is, in fact, what takes place.

It is, perhaps, worth passing observation that, from the nutritive dependence of the animal upon the vegetable,

and the fact that a conversion of the energy of the one to the purposes of the other cannot occur without loss, the mean energy absorbed daily by the vegetable for the purpose of growth must greatly exceed that used in animal growth; so that the chemical potential energy of vegetation upon the earth is much greater than the energy of all kinds represented in the animal configurations.[1] It appears, too, that in the power possessed by the vegetable of remaining comparatively inactive, of surviving hard times by the expenditure and absorption of but little, the vegetable constitutes a veritable reservoir for the uniform supply of the more unstable and active animal.

Finally, on the question of the manner of origin of organic systems, it is to be observed that, while the life of the present is very surely the survival of the fittest of the tendencies and chances of the past, yet, in the initiation of the organised world, a single chance may have decided a whole course of events: for, once originated, its own law secures its increase, although within the new order of actions, the law of the fittest must assert itself. That such a progressive material system as an organism was possible, and at some remote period was initiated, is matter of knowledge; whether or not the initiatory living configuration was rare and fortuitous, or the probable result of the general action of physical laws acting among innumerable chances, must remain matter of

[1] I find a similar conclusion arrived at in Semper's _Animal Life_, p. 52.

speculation. In the event of the former being the truth, it is evidently possible, in spite of a large finite number of habitable worlds, that life is non-existent elsewhere. If the latter is the truth, it is almost certain that there is life in all, or many of those worlds.

EVOLUTION AND ACCELERATION OF ACTIVITY

The primary factor in evolution is the "struggle for existence." This involves a "natural selection" among the many variations of the organism. If we seek the underlying causes of the struggle, we find that the necessity of food and (in a lesser degree) the desire for a mate are the principal causes of contention. The former is much the more important factor, and, accordingly, we find the greater degree of specialisation based upon it.

The present view assumes a dynamic necessity for its demands involved in the nature of the organism as such. This assumption is based on observation of the outcome of its unconstrained growth, reproduction, and life-acts. We have the same right to assert this of the organism as we have to assert that retardation and degradation attend the actions of inanimate machines, which assertion, also, is based on observation of results. Thus we pass from the superficial statements that organisms require food in order to live, or that organisms desire food, to the more fundamental one that:

The organism is a configuration of matter which absorbs energy acceleratively, without limit, when unconstrained.

This is the dynamic basis for a "struggle for existence." The organism being a material system responding to accession of energy with fresh demands, and energy being limited in amount, the struggle follows as a necessity. Thus, evolution guiding' the steps of the energy-seeking organism, must presuppose and find its origin in that inherent property of the organism which determines its attitude in presence of available energy.

Turning to the factor, "adaptation," we find that this also must presuppose, in order to be explicable, some quality of aggressiveness on the part of the organism. For adaptation in this or that direction is the result of repulse or victory, and, therefore, we must presuppose an attack. The attack is made by the organism in obedience to its law of demand; we see in the adaptation of the organism but the accumulated wisdom derived from past defeats and victories.

Where the environment is active, that is living, adaptation occurs on both sides. Improved means of defence or improved means of attack, both presuppose activity. Thus the reactions to the environment, animate and inanimate, are at once the outcome of the eternal aggressiveness of the organism, and the source of fresh aggressiveness upon the resources of the medium.

As concerns the "survival of the fittest" (or "natural selection"), we can, I think, at once conclude that the organism which best fulfils the organic law under the circumstances of supply is the "fittest," _ipso facto._ In many

cases this is contained in the commonsense consideration, that to be strong, consistent with concealment from enemies which are stronger, is best, as giving the organism mastery over foes which are weaker, and generally renders it better able to secure supplies. Weismann points out that natural selection favours early and abundant reproduction. But whether the qualifications of the "fittest" be strength, fertility, cunning, fleetness, imitation, or concealment, we are safe in concluding that growth and reproduction must be the primary qualities which at once determine selection and are fostered by it. Inherent in the nature of the organism is accelerated absorption of energy, but the qualifications of the "fittest" are various, for the supply of energy is limited, and there are many competitors for it. To secure that none be wasted is ultimately the object of natural selection, deciding among the eager competitors what is best for each.

In short, the facts and generalisations concerning evolution must presuppose an organism endowed with the quality of progressive absorption of energy, and retentive of it. The continuity of organic activity in a world where supplies are intermittent is evidently only possible upon the latter condition. Thus it appears that the dynamic attitude of the organism, considered in these pages, occupies a fundamental position regarding its evolution.

We turn to the consideration of old age and death, endeavouring to discover in what relation they stand to the innate progressiveness of the organism.

THE PERIODICITY OF THE ORGANISM AND THE LAW OF PROGRESSIVE ACTIVITY

The organic system is essentially unstable. Its aggressive attitude is involved in the phenomenon of growth, and in reproduction which is a form of growth. But the energy absorbed is not only spent in growth. It partly goes, also, to make good the decay which arises from the instability of the organic unit. The cell is molecularly perishable. It possesses its entity much as a top keeps erect, by the continual inflow of energy. Metabolism is always taking place within it. Any other condition would, probably, involve the difficulties of perpetual motion.

The phenomenon of old age is not evident in the case of the unicellular organism reproducing by fission. At any stage of its history all the individuals are of the same age: all contain a like portion of the original cell, so far as this can be regarded as persisting where there is continual flux of matter and energy. In the higher organisms death is universally evident. Why is this?

The question is one of great complexity. Considered from the more fundamental molecular point of view we should perhaps look to failure of the power of cell division as the condition of mortality. For it is to this phenomenon—that of cell division—that the continued life of the protozoon is to be ascribed, as we have already seen. Reproduction is, in fact, the saving factor here.

As we do not know the source or nature of the stimulus

responsible for cell division we cannot give a molecular account of death in the higher organisms. However we shall now see that, philosophically, we are entitled to consider reproduction as a saving factor in this case also; and to regard the death of the individual much as we regard the fall of the leaf from the tree: _i.e._ as the cessation of an outgrowth from a development extending from the past into the future. The phenomena of old age and natural death are, in short, not at variance with the progressive activity of the organism. We perceive this when we come to consider death from the evolutionary point of view.

Professor Weismann, in his two essays, "The Duration of Life," and "Life and Death,"[1] adopts and defends the view that "death is not a primary necessity but that it has been secondarily acquired by adaptation." The cell was not inherently limited in its number of cell-generations. The low unicellular organisms are potentially immortal, the higher multicellular forms with well-differentiated organs contain the germs of death within themselves.

He finds the necessity of death in its utility to the species. Long life is a useless luxury. Early and abundant reproduction is best for the species. An immortal individual would gradually become injured and would be valueless or even harmful to the species by taking the place of those that are sound. Hence natural selection will shorten life.

[1] See his _Biological Memoirs._ Oxford, 1888.

Weismann contends against the transmission of acquired characters as being unproved.[1] He bases the appearance of death on variations in the reproductive cells, encouraged by the ceaseless action of natural selection, which led to a differentiation into perishable somatic cells and immortal reproductive cells. The time-limit of any particular organism ultimately depends upon the number of somatic cell-generations and the duration of each generation. These quantities are "predestined in the germ itself" which gives rise to each individual. "The existence of immortal metazoan organisms is conceivable," but their capacity for existence is influenced by conditions of the external world; this renders necessary the process of adaptation. In fact, in the differentiation of somatic from reproductive cells, material was provided upon which natural selection could operate to shorten or to lengthen the life of the individual in accordance with the needs of the species. The soma is in a sense "a secondary appendage of the real bearer of life—the reproductive cells." The somatic cells probably lost their immortal qualities, on this immortality becoming useless to the species. Their mortality may have been a mere consequence of their differentiation (loc. cit., p. 140), itself due to natural selection. "Natural death was not," in fact, "introduced from absolute intrinsic necessity inherent in the nature of living matter, but on grounds of utility,

[1] Biological Memoirs, p. 142.

that is from necessities which sprang up, not from the general conditions of life, but from those special conditions which dominate the life of multicellular organisms."

On the inherent immortality of life, Weismann finally states: "Reproduction is, in truth, an essential attribute of living matter, just as the growth which gives rise to it.... Life is continuous, and not periodically interrupted: ever since its first appearance upon the Earth in the lowest organism, it has continued without break; the forms in which it is manifest have alone undergone change. Every individual alive today—even the highest—is to be derived in an unbroken line from the first and lowest forms." [1]

At the present day the view is very prevalent that the soma of higher organisms is, in a sense, but the carrier for a period of the immortal reproductive cells (Ray Lankester)[2]—an appendage due to adaptation, concerned in their supply, protection, and transmission. And whether we regard the time-limit of its functions as due to external constraints, recurrently acting till their effects become hereditary, or to variations more directly of internal origin, encouraged by natural selection, we see in old age and death phenomena ultimately brought about in obedience to the action of an environment. These are not inherent in the properties of living matter. But, in spite

[1] Loc. cit., p. 159

[2] Geddes and Thomson, The Evolution of Sex, chap. xviii.

of its mortality, the body remains a striking manifestation of the progressiveness of the organism, for to this it must be ascribed. To it energy is available which is denied to the protozoon. Ingenious adaptations to environment are more especially its privilege. A higher manifestation, however, was possible, and was found in the development of mind. This, too, is a servant of the cell, as the genii of the lamp. Through it energy is available which is denied to the body. This is the masterpiece of the cell. Its activity dates, as it were, but from yesterday, and today it inherits the most diverse energies of the Earth.

Taking this view of organic succession, we may liken the individual to a particle vibrating for a moment and then coming to rest, but sweeping out in its motion one wave in the continuous organic vibration travelling from the past into the future. But as this vibration is one spreading with increased energy from each vibrating particle, its propagation involves a continual accelerated inflow of energy from the surrounding medium, a dynamic condition unknown in periodic effects transmitted by inanimate actions, and, indeed, marking the fundamental difference between the dynamic attitudes of the animate and inanimate.

We can trace the periodic succession of individuals on a diagram of activity with some advantage. Considering, first, the case of the unicellular organism reproducing by subdivision and recalling that conditions, definite and inevitable, oppose a limit to the rate of growth, or, for our

present purpose, rate of consumption of energy, wc proceed as follows:

{Fig. 1}

Along a horizontal axis units of time are measured; along a vertical axis units of energy. Then the life-history of the amoeba, for example, appears as a line such as A in Fig. 1. During the earlier stages of its growth the rate of absorption of energy is small; so that in the unit interval of time, t, the small quantity of energy, e_1, is absorbed. As life advances, the activity of the organism augments, till finally this rate attains a maximum, when e_2 units of energy are consumed in the unit of time.[1]

[1] Reference to p. 76, where the organic system is treated as purely mechanical, may help readers to understand what is involved in this curve. The solar engine may, unquestionably, have its activity defined by such a curve. The organism is, indeed, more complex; but neither this fact nor our ignorance of its mechanism, affects the principles which justify the diagram.

On this diagram reproduction, on the part of the organism, is represented by a line which repeats the curvature of the parent organism originating at such a point as P in the path of the latter, when the rate of consumption of energy has become constant. The organism A has now ceased to act as a unit. The products of fission each carry on the vital development of

{Fig. 2}

the species along the curve B, which may be numbered (2), to signify that it represents the activity of two individuals, and so on, the numbering advancing in geometrical progression. The particular curvature adopted in the diagram is, of course, imaginary; but it is not of an indeterminate nature. Its course for any species is a characteristic of fundamental physical importance, regarding the part played in nature by the particular organism.

In Fig. 2 is represented the path of a primitive multicellular organism before the effects of competition produced or fostered its mortality. The lettering of Fig. 1 applies; the successive reproductive acts are marked P1, P2; Q1, Q2, etc., in the paths of the successive individuals.

{Fig. 3}

The next figure (Fig. 3) diagrammatically illustrates death in organic history. The path ever turns more and more from the axis of energy, till at length the point is reached when no more energy is available; a tangent to the curve at this point is at right angles to the axis of energy and parallel to the time axis. The death point is reached, and however great a length we measure along the axis of time, no further consumption of energy is

indicated by the path of the organism. Drawing the line beyond the death point is meaningless for our present purpose.

It is observable that while the progress of animate nature finds its representation on this diagram by lines sloping _upwards_ from left to right, the course of events in inanimate nature—for example, the history of the organic configuration after death, or

{Fig. 4}

the changes progressing—let us say, in the solar system, or in the process of a crystallisation, would appear as lines sloping downwards from left to right.

Whatever our views on the origin of death may be, we have to recognise a periodicity of functions in the life-history of the successive individuals of the present day; and whether or not we trace this directly or indirectly to

a sort of interference with the rising wave of life, imposed by the activity of a series of derived units, each seeking energy, and in virtue of its adaptation each being more fitted to obtain it than its predecessor, or even leave the idea of interference out of account altogether in the origination or perpetuation of death, the truth of the diagram (Fig. 4) holds in so far as it may be supposed to graphically represent the dynamic history of the individual. The point chosen on the curve for the origination of a derived unit is only applicable to certain organisms, many reproducing at the very close of life. A chain of units are supposed here represented.[1]

THE LENGTH OF LIFE

If we lay out waves as above to a common scale of time for different species, the difference of longevity is shown in the greater or less number of vibrations executed in a given time, _i.e._ in greater or less "frequency." We cannot indeed draw the curvature correctly, for this would necessitate a knowledge which we have not of the activity of the organism at different periods of its life-history, and so neither can we plot the direction of the organic line of propagation with respect to the

[1] Projecting upon the axes of time and energy any one complete vibration, as in Fig. 4, the total energy consumed by the organism during life is the length E on the axis of energy, and its period of life is the length T on the time-axis. The mean activity is the quotient E/T.

axes of reference as this involves a knowledge of the mean activity.[1]

The group of curves which follow, relating to typical animals possessing very different activities (Fig. 5), are therefore entirely diagrammatic, except in respect to the approximate

{Fig. 5}

longevity of the organisms. (1) might represent an animal of the length of life and of the activity of Man; (2), on the same scale of longevity,

[1] In the relative food-supply at various periods of life the curvature is approximately determinable.

one of the smaller mammals; and (3), the life-history of a cold blooded animal living to a great age; e.g. certain of the reptilia.

It is probable, that to conditions of structural development, under the influence of natural selection, the question of longer or shorter life is in a great degree referable. Thus, development along lines of large growth will tend to a slow rate of reproduction from the simple fact that unlimited energy to supply abundant reproduction is not procurable, whatever we may assume as to the strength or cunning exerted by the individual in its efforts to obtain its supplies. On the other hand, development along lines of small growth, in that reproduction is less costly, will probably lead to increased rate of reproduction. It is, in fact, matter of general observation that in the case of larger animals the rate of reproduction is generally slower than in the case of smaller animals. But the rate of reproduction might be expected to have an important influence in determining the particular periodicity of the organism. Were we to depict in the last diagram, on the same time-scale as Man, the vibrations of the smaller and shorter-lived living things, we would see but a straight line, save for secular variations in activity, representing the progress of the species in time: the tiny thrills of its units lost in comparison with the yet brief period of Man.

The interdependence of the rate of reproduction and

the duration of the individual is, indeed, very probably revealed in the fact that short-lived animals most generally reproduce themselves rapidly and in great abundance, and vice versa. In many cases where this appears contradicted, it will be found that the young are exposed to such dangers that but few survive (_e.g._ many of the reptilia, etc.), and so the rate of reproduction is actually slow.

Death through the periodic rigour of the inanimate environment calls forth phenomena very different from death introduced or favoured by competition. A multiplicity of effects simulative of death occur. Organisms will, for example, learn to meet very rigorous conditions if slowly introduced, and not permanent. A transitory period of want can be tided over by contrivance. The lily withdrawing its vital forces into the bulb, protected from the greatest extremity of rigour by seclusion in the Earth; the trance of the hibernating animal; are instances of such contrivances.

But there are organisms whose life-wave truly takes up the periodicity of the Earth in its orbit. Thus the smaller animals and plants, possessing less resources in themselves, die at the approach of winter, propagating themselves by units which, whether egg or seed, undergo a period of quiescence during the season of want. In these quiescent units the energy of the organism is potential, and the time-energy function is in abeyance. This condition is, perhaps, foreshadowed in the encystment of the amoeba in resistance to drought. In most cases of hibernation the time-energy function seems maintained at a loss of potential by the organism, a diminished vital consumption of energy being carried on at the expense of the stored energy of the tissues. So, too, even among the largest organisms there will be a diminution of activity periodically inspired by climatological conditions. Thus, wholly or in part, the activity of organisms is recurrently affected by the great energy—tides set up by the Earth's orbital motion.

{Fig. 6}

Similarly in the phenomenon of sleep the organism responds to the Earth's axial periodicity, for in the interval of night a period of impoverishment has to be endured. Thus the diurnal waves of energy also meet a response in the organism. These tides and waves of activity would appear as larger and smaller ripples

on the life-curve of the organism. But in some, in which life and death are encompassed in a day, this would not be so; and for the annual among plants, the seed rest divides the waves with lines of no activity (Fig. 6).

Thus, finally, we regard the organism as a dynamic phenomenon passing through periodic variations of intensity. The material systems concerned in the transfer of the energy rise, flourish, and fall in endless succession, like cities of ancient dynasties. At points of similar phase upon the waves the rate of consumption of energy is approximately the same; the functions, too, which demand and expend the energy are of similar nature.

That the rhythm of these events is ultimately based on harmony in the configuration and motion of the molecules within the germ seems an unavoidable conclusion. In the life of the individual rhythmic dynamic phenomena reappear which in some cases have no longer a parallel in the external world, or under conditions when the individual is no longer influenced by these external conditions.,, In many cases the periodic phenomena ultimately die out under new influences, like the oscillations of a body in a viscous medium; in others when they seem to be more deeply rooted in physiological conditions they persist.

The "length of life is dependent upon the number

[1] The _Descent of Man._

of generations of somatic cells which can succeed one another in the course of a single life, and furthermore the number as well as the duration of each single cell-generation is predestined in the germ itself."[1]

Only in the vague conception of a harmonising or formative structural influence derived from the germ, perishing in each cell from internal causes, but handed from cell to cell till the formative influence itself degrades into molecular discords, does it seem possible to form any physical representation of the successive events of life. The degradation of the molecular formative influence might be supposed involved in its frequent transference according to some such dynamic actions as occur in inanimate nature. Thus, ultimately, to the waste within the cell, to the presence of a force retardative of its perpetual harmonic motions, the death of the individual is to be ascribed. Perhaps in protoplasmic waste the existence of a universal death should be recognised. It is here we seem to touch inanimate nature; and we are led back to a former conclusion that the organism in its unconstrained state is to be regarded as a contrivance for evading the dynamic tendencies of actions in which lifeless matter participates.[2]

[1] Weismann, _Life and Death; Biological Memoirs_, p. 146.

[2] In connection with the predestinating power and possible complexity of the germ, it is instructive to reflect on the very great molecular population of even the smallest spores—giving rise to very simple forms. Thus, the spores of the unicellular Schizomycetes are estimated to dimensions as low as 1/10,000 of a millimetre in diameter (Cornil et Babes, _Les Batteries_, 1. 37). From Lord Kelvin's estimate of the number of molecules in water, comprised within the length of a wave-length of yellow light (_The Size of Atoms_, Proc. R. I., vol. x., p. 185) it is probable that such spores contain some 500,000 molecules, while one hundred molecules range along a diameter.

THE NUMERICAL ABUNDANCE OF LIFE

We began by seeking in various manifestations of life a dynamic principle sufficiently comprehensive to embrace its very various phenomena. This, to all appearance, found, we have been led to regard life, to a great extent, as a periodic dynamic phenomenon. Fundamentally, in that characteristic of the contrivance, which leads it to respond favourably to transfer of energy, its enormous extension is due. It is probable that to its instability its numerical abundance is to be traced—for this, necessitating the continual supply of all the parts already formed, renders large, undifferentiated growth, incompatible with the limited supplies of the environment. These are fundamental conditions of abundant life upon the Earth.

Although we recognise in the instability of living systems the underlying reason for their numerical abundance, secondary evolutionary causes are at work. The most important of these is the self-favouring nature of the phenomenon of reproduction. Thus there is a tendency not only to favour reproductiveness, but early reproductiveness, in the form of one prolific reproductive.

act, after which the individual dies.[1] Hence the wavelength of the species diminishes, reproduction is more frequent, and correspondingly greater numbers come and go in an interval of time.

Another cause of the numerical abundance of life exists, as already stated, in the conditions of nourishment. Energy is more readily conveyed to the various parts of the smaller mass, and hence the lesser organisms will more actively functionate; and this, as being the urging dynamic attitude, as well as that most generally favourable in the struggle, will multiply and favour such forms of life. On the other hand, however, these forms will have less resource within themselves, and less power of endurance, so that they are only suitable to fairly uniform conditions of supply; they cannot survive the long continued want of winter, and so we have the seasonal abundance of summer. Only the larger and more resistant organisms, whether animal or vegetable, will, in general, populate the Earth from year to year. From this we may conclude that, but for the seasonal energy-tides, the development of life upon the globe had gone along very different lines from those actually followed. It is, indeed, possible that the evolution of the larger organisms would not have occurred; there would have been no vacant place for their development, and a being so endowed as Man could hardly

[1] Weismann, _The Duration of Life._

have been evolved. We may, too, apply this reasoning elsewhere, and regard as highly probable, that in worlds which are without seasonal influences, the higher developments of life have not appeared; except they have been evolved under other conditions, when they might for a period persist. We have, indeed, only to picture to ourselves what the consequence of a continuance of summer would be on insect life to arrive at an idea of the antagonistic influences obtaining in such worlds to the survival of larger organisms.

It appears that to the dynamic attitude of life in the first place, and secondarily to the environmental conditions limiting undifferentiated growth, as well as to the action of heredity in transmitting the reproductive qualities of the parent to the offspring, the multitudes of the pines, and the hosts of ants, are to be ascribed. Other causes are very certainly at work, but these, I think, must remain primary causes.

We well know that the abundance of the ants and pines is not a tithe of the abundance around us visible and invisible. It is a vain endeavour to realise the countless numbers of our fellow-citizens upon the Earth; but, for our purpose, the restless ants, and the pines solemnly quiet in the sunshine, have served as types of animate things. In the pine the gates of the organic have been thrown open that the vivifying river of energy may flow in. The ants and the butterflies sip for a brief moment of its waters, and again vanish into the

inorganic: life, love and death encompassed in a day.

Whether the organism stands at rest and life comes to it on the material currents of the winds and waters, or in the vibratory energy of the æther; or, again, whether with restless craving it hurries hither and thither in search of it, matters nothing. The one principle—the accelerative law which is the law of the organic—urges all alike onward to development, reproduction and death. But although the individual dies death is not the end; for life is a rhythmic phenomenon. Through the passing ages the waves of life persist: waves which change in their form and in the frequency to which they are attuned from one geologic period to the next, but which still ever persist and still ever increase. And in the end the organism outlasts the generations of the hills.

THE BRIGHT COLOURS OF ALPINE FLOWERS [1]

IT is admitted by all observers that many species of flowering plants growing on the higher alps of mountainous regions display a more vivid and richer colour in their bloom than is displayed in the same species growing in the valleys. That this is actually the case, and not merely an effect produced upon the observer by the scant foliage rendering the bloom more conspicuous, has been shown by comparative microscopic examination of the petals of species growing on the heights and in the valleys. Such examination has revealed that in many cases pigment granules are more numerous in the individuals growing at the higher altitudes. The difference is specially marked in Myosotis sylvatica, Campanula rotundifolia, Ranunculus sylvaticus, Galium cruciatum, and others. It is less marked in the case of Thymus serpyllum and Geranium sylvaticum; while in Rosa alpina and Erigeron alpinus no difference is observable.[2]

In the following cases a difference of intensity of colour is, according to Kerner ("Pflanzenleben," 11. 504), especially noticeable:— _Agrostemma githago, Campanula

[1] _Proc. Royal Dublin Society_, 1893.

[2] G. Bonnier, quoted by De Varigny, _Experimental Evolution_, p. 55.

pusilla, Dianthus inodorus (silvestris), Gypsophila repens, Lotus corniculatus, Saponaria ocymoides, Satureja hortensis, Taraxacumm officinale, Vicia cracca, and Vicia sepium._

To my own observation this beautiful phenomenon has always appeared most obvious and impressive. It appears to have struck many unprofessional observers. Helmholtz offers the explanation that the vivid colours are the result of the brighter sunlight of the heights. It has been said, too, that they are the direct chemical effects of a more highly ozonized atmosphere. The latter explanation I am unable to refer to its author. The following pages contain a suggestion on the matter, which occurred to me while touring, along with Henry H. Dixon, in the Linthal district of Switzerland last summer.[1]

If the bloom of these higher alpine flowers is especially pleasing to our own œsthetic instincts, and markedly conspicuous to us as observers, why not also especially attractive and conspicuous to the insect whose mission it is to wander from flower to flower over the pastures? The answer to this question involves the hypothesis I would advance as accounting for the bright colours of high-growing individuals. In short, I believe a satisfactory explanation is to be found in the conditions of insect life in the higher alps.

In the higher pastures the summer begins late and

[1] The summer of 1892.

closes early, and even in the middle of summer the day closes in with extreme cold, and the cold of night is only dispelled when the sun is well up. Again, clouds cover the heights when all is clear below, and cold winds sweep over them when there is warmth and shelter in the valleys. With these rigorous conditions the pollinating insects have to contend in their search for food, and that when the rival attractions of the valleys below are so many. I believe it is these rigorous conditions which are indirectly responsible for the bright colours of alpine flowers. For such conditions will bring about a comparative scarcity of insect activity on the heights; and a scarcity or uncertainty in the action of insect agency in effecting fertilization will intensify the competition to attract attention, and only the brightest blooms will be fertilized.[1]

This will be a natural selection of the brightest, or the

[1] Grant Allen, I have recently learned, advances in _Science in Arcady_ the theory that there is a natural selective cause fostering the bright blooms of alpines. The selective cause is, however, by him referred to the greater abundance of butterfly relatively to bee fertilizers. The former, he says, display more æsthetic instinct than bees. In the valley the bees secure the fertilization of all. I may observe that upon the Fridolins Alp all the fertilizers we observed were bees. I have always found butterflies very scarce at altitudes of 7,000 to 8,000 feet. The alpine bees are very light in body, like our hive bee, and I do not think rarefaction of the atmosphere can operate to hinder its ascent to the heights, as Grant Allen suggests. The observations on the death-rate of bees and butterflies on the glacier, to be referred to presently, seem to negative such a hypothesis, and to show that a large preponderance of bees over butterflies make their way to the heights.

brightest will be the fittest, and this condition, along with the influence of heredity, will encourage a race of vivid flowers. On the other hand, the more scant and uncertain root supply, and the severe atmospheric conditions, will not encourage the grosser struggle for existence which in the valleys is carried on so eagerly between leaves and branches—the normal offensive and defensive weapons of the plant—and so the struggle becomes refined into the more æsthetic one of colour and brightness between flower and flower. Hence the scant foliage and vivid bloom would be at once the result of a necessary economy, and a resort to the best method of securing reproduction under the circumstances of insect fertilizing agency. Or, in other words, while the luxuriant growth is forbidden by the conditions, and thus methods of offence and defence, based upon vigorous development, reduced in importance, it would appear that the struggle is mainly referred to rivalry for insect preference. It is probable that this is the more economical manner of carrying on the contest.

In the valleys we see on every side the struggle between the vegetative organs of the plant; the soundless battle among the leaves and branches. The blossom here is carried aloft on a slender stem, or else, taking but a secondary part in the contest, it is relegated to obscurity (P1. XII.). Further up on the mountains, where the conditions are more severe and the supplies less abundant, the leaf and branch assume lesser dimensions, for they are costly weapons to provide and the elements are unfriendly

to their existence (Pl. XIII.). Still higher, approaching the climatic limit of vegetable life, the struggle for existence is mainly carried on by the æsthetic rivalry of lowly but conspicuous blossoms.

As regards the conditions of insect life in the higher alps, it came to my notice in a very striking manner that vast numbers of such bees and butterflies as venture up perish in the cold of night time. It appears as if at the approach of dusk these are attracted by the gleam of the snow, and quitting the pastures, lose themselves upon the glaciers and firns, there to die in hundreds. Thus in an ascent of the Tödi from the Fridolinshüte we counted in the early dawn sixty-seven frozen bees, twenty-nine dead butterflies, and some half-dozen moths on the Biferten Glacier and Firn. These numbers, it is to be remembered, only included those lying to either side of our way over the snow, so that the number must have mounted up to thousands when integrated over the entire glacier and firn. Approaching the summit none were found. The bees resembled our hive bee in appearance, the butterflies resembled the small white variety common in our gardens, which has yellow and black upon its wings. One large moth, striped across the abdomen, and measuring nearly two inches in length of body, was found. Upon our return, long after the sun's rays had grown strong, we observed some of the butterflies showed signs of reanimation. We descended so quickly to avoid the inconvenience of the soft snow that we had time for no

close observation on the frozen bees. But dead bees are common objects upon the snows of the alps.

These remarks I noted down roughly while at Linthal last summer, but quite recently I read in Natural Science[1] the following note:

"Late Flowering Plants.—While we write, the ivy is in flower, and bees, wasps, and flies are jostling each other and struggling to find standing-room on the sweet-smelling plant. How great must be the advantage obtained by this plant through its exceptional habit of flowering in the late autumn, and ripening its fruit in the spring. To anyone who has watched the struggle to approach the ivy-blossom at a time when nearly all other plants are bare, it is evident that, as far as transport of pollen and cross-fertilization go, the plant could not flower at a more suitable time. The season is so late that most other plants are out of flower, but yet it is not too late for many insects to be brought out by each sunny day, and each insect, judging by its behaviour, must be exceptionally hungry.

"Not only has the ivy the world to itself during its flowering season, but it delays to ripen its seed till the spring, a time when most other plants have shed their seed, and most edible fruits have been picked by the birds. Thus birds wanting fruit in the spring can obtain little but ivy, and how they appreciate the ivy berry is evident

[1] For December, 1892, vol. i., p. 730.

by the purple stains everywhere visible within a short distance of the bush."

These remarks suggest that the ivy adopts the converse attitude towards its visitors to that forced upon the alpine flower. The ivy bloom is small and inconspicuous, but then it has the season to itself, and its inconspicuousness is no disadvantage, _i.e._ if one plant was more conspicuous than its neighbours, it would not have any decided advantage where the pollinating insect is abundant and otherwise unprovided for. Its dark-green berries in spring, which I would describe as very inconspicuous, have a similar advantage in relation to the necessities of bird life.

The experiments of M. C. Flahault must be noticed. This naturalist grew seeds of coloured flowers which had ripened in Paris, part in Upsala, and part in Paris; and seed which had ripened in Upsala, part at Paris, and part at Upsala. The flowers opening in the more northern city were in most cases the brighter.[1] If this observation may be considered indisputable, as appears to be the case, the question arises, Are we to regard this as a direct effect of the more rigorous climate upon the development of colouring matter on the blooms opening at Upsala? If we suppose an affirmative answer, the theory of direct effect by sun brightness must I think be abandoned. But I venture to think that the explanation of the Upsala

[1] Quoted by De Varigny, _Experimental Evolution_, p. 56.

experiment is not to be found in direct climatic influence upon the colour, but in causes which lie deeper, and involve some factors deducible from biological theory.

The organism, as a result of the great facts of heredity and of the survival of the fittest, is necessarily a system which gathers experience with successive generations; and the principal lesson ever being impressed upon it by external events is economy. Its success depends upon the use it makes of its opportunities for the reception of energy and the economy attained in disposing of what is gained.

With regard to using the passing opportunity the entire seasonal development of life is a manifestation of this attitude, and the fleetness, agility, etc., of higher organisms are developments in this direction. The higher vegetable organism is not locomotory, save in the transferences of pollen and seed, for its food comes to it, and the necessary relative motion between food and organism is preserved in the quick motion of radiated energy from the sun and the slower motion of the winds on the surface of the earth. But, even so, the vegetable organism must stand ever ready and waiting for its supplies. Its molecular parts must be ready to seize the prey offered to it, somewhat as the waiting spider the fly. Hence, the plant stands ready; and every cloud with moving shadow crossing the fields handicaps the shaded to the benefit of the unshaded plant in the adjoining field. The open bloom

is a manifestation of the generally expectant attitude of the plant, but in relation to reproduction.

As regards economy, any principle of maximum economy, where many functions have to be fulfilled, will, we may very safely predict, involve as far as possible mutual helpfulness in the processes going on. Thus the process of the development towards meeting any particular external conditions, A, suppose, will, if possible, tend to forward the development towards meeting conditions B; so that, in short, where circumstances of morphology and physiology are favourable, the ideally economical system will be attained when in place of two separate processes, a, ß, the one process y, cheaper than a + ß, suffices to advance development simultaneously in both the directions A and B. The economy is as obvious as that involved in "killing two birds with the one stone"—if so crude a simile is permissible—and it is to be expected that to foster such economy will be the tendency of evolution in all organic systems subjected to restraints as those we are acquainted with invariably are.

Such economy might be simply illustrated by considering the case of a reservoir of water elevated above two hydraulic motors, so that the elevated mass of water possessed gravitational potential. The available energy here represents the stored-up energy in the organism. How best may the water be conveyed to the two motors [the organic systems reacting towards conditions A and B] so

that as little energy as possible is lost in transit? If the motors are near together it is most economical to use the one conduit, which will distribute the requisite supply of water to both. If the motors are located far asunder it will be most economical to lay separate conduits. There is greatest economy in meeting a plurality of functions by the same train of physiological processes where this is consistent with meeting other demands necessitated by external or internal conditions.

But an important and obvious consequence arises in the supply of the two motors from the one conduit. We cannot work one motor without working the other. If we open a valve in the conduit both motors start into motion and begin consuming the energy stored in the tank. And although they may both under one set of conditions be doing useful and necessary work, in some other set of conditions it may be needless for both to be driven.

This last fact is an illustration of a consideration which must enter into the phenomenon which an eminent biologist speaks of as physiological or unconscious "memory,"[1] For the development of the organism from the ovum is but the starting of a train of interdependent events of a complexity depending upon the experience of the past.

[1] Ewald Hering, quoted by Ray Lankaster, _The Advancement of Science_, p. 283.

In short, we may suppose the entire development of the plant, towards meeting certain groups of external conditions, physiologically knit together according as Nature tends to associate certain groups of conditions. Thus, in the case in point, climatic rigour and scarcity of pollinating agency will ever be associated; and in the long experience of the past the most economical physiological attitude towards both is, we may suppose, adopted; so that the presence of one condition excites the apparent unconscious memory of the other. In reality the process of meeting the one condition involves the process and development for meeting the other.

And this consideration may be extended very generally to such organisms as can survive under the same associated natural conditions, for the history of evolution is so long, and the power of locomotion so essential to the organism at some period in its life history, that we cannot philosophically assume a local history for members of a species even if widely severed geographically at the present day. At some period in the past then, it is very possible that the individuals today thriving at Paris, acquired the experience called out at Upsala. The perfection of physiological memory inspires no limit to the date at which this may have occurred—possibly the result of a succession of severe seasons at Paris; possibly the result of migrations —and the seed of many flowering plants possess means of migration only inferior to those possessed by the flying and swimming animals. But, again, possibly the experi-

ence was acquired far back in the evolutionary history of the flower.[1]

But a further consideration arises. Not only at each moment in the life of the individual must maximum income and most judicious expenditure be considered, but in its whole life history, and even over the history of its race, the efficiency must tend to be a maximum. This principle is even carried so far that when necessary it leads to the death of the individual, as in the case of those organisms which, having accomplished the reproductive act, almost immediately expire. This view of nature may be repellent, but it is, nevertheless, evident that we are parts of a system which ruthlessly sacrifices the individual on general grounds of economy. Thus, if the curve which defines the mean rate of reception of energy of all kinds at different periods in the life of the organism be opposed by a second curve, drawn below the axis along which time is measured, representing the mean rate of expenditure of energy on development, reproduction, etc. (Fig. 7), this latter curve, which is, of course,

[1] The blooms of self-fertilising, and especially of cleistogamic plants (_e.g._ Viola), are examples of unconscious memory, or unconscious "association of ideas" leading to the development of organs now functionless. The _Pontederia crassipes_ of the Amazon, which develops its floating bladders when grown in water, but aborts them rapidly when grown on land, and seems to retain this power of adaptation to the environment for an indefinite period of time, must act in each case upon an unconscious memory based upon past experience. Many other cases might be cited.

physiologically dependent on the former, must be of such a nature from its origin to its completion in death, that the condition is realized of the most economical rate of expenditure at each period of life.[1] The rate of expenditure of energy at any period of life is, of course, in such a curve defined by the slope of the curve towards the axis of time at the period in question; but this particular slope _must be led to by a previous part of the curve, and involves its past and future course to a very great extent_.

{Fig. 7}

There will, therefore, be impressed upon the organism by the factors of evolution a unified course of economical expenditure completed only by its death, and which will give to the developmental progress of the individual its prophetic character.

In this way we look to the unified career of each organic unit, from its commencement in the ovum to the day

[1] See _The Abundance of Life_.

when it is done with vitality, for that preparation for momentous organic events which is in progress throughout the entire course of development; and to the economy involved in the welding of physiological processes for the phenomenon of physiological memory, wherein we see reflected, as it were, in the development of the organism, the association of inorganic restraints occurring in nature which at some previous period impressed itself upon the plastic organism. We may picture the seedling at Upsala, swayed by organic memory and the inherited tendency to an economical preparation for future events, gradually developing towards the æsthetic climax of its career. In some such manner only does it appear possible to account for the prophetic development of organisms, not alone to be observed in the alpine flowers, but throughout nature.

And thus, finally, to the effects of natural selection and to actions defined by general principles involved in biology, I would refer for explanation of the manner in which flowers on the Alps develop their peculiar beauty.

MOUNTAIN GENESIS

OUR ancestors regarded mountainous regions with feelings of horror, mingled with commiseration for those whom an unkindly destiny had condemned to dwell therein. We, on the other hand, find in the contemplation of the great alps of the Earth such peaceful and elevated thoughts, and such rest to our souls, that it is to those very solitudes we turn to heal the wounds of ife. It is difficult to explain the cause of this very different point of view. It is probably, in part, to be referred to that cloud of superstitious horror which, throughout the Middle Ages, peopled the solitudes with unknown terrors; and, in part, to the asceticism which led the pious to regard the beauty and joy of life as snares to the soul's well-being. In those eternal solitudes where the overwhelming forces of Nature are most in evidence, an evil principle must dwell or a dragon's dreadful brood must find a home.

But while in our time the aesthetic aspect of the hills appeals to all, there remains in the physical history of the mountains much that is lost to those who have not shared in the scientific studies of alpine structure and genesis. They lose a past history which for interest com-

petes with anything science has to tell of the changes of the Earth.

Great as are the physical features of the mountains compared with the works of Man, and great as are the forces involved compared with those we can originate or control, the loftiest ranges are small contrasted with the dimensions of the Earth. It is well to bear this in mind. I give here (Pl. XV.) a measured drawing showing a sector cut from a sphere of 50 cms. radius; so much of it as to exhibit the convergence of its radial boundaries which if prolonged will meet at the centre. On the same scale as the radius the diagram shows the highest mountains and the deepest ocean. The average height of the land and the average depth of the ocean are also exhibited. We see how small a movement of the crust the loftiest elevation of the Himalaya represents and what a little depression holds the ocean.

Nevertheless, it is not by any means easy to explain the genesis of those small elevations and depressions. It would lead us far from our immediate subject to discuss the various theoretical views which have been advanced to account for the facts. The idea that mountain folds, and the lesser rugosities of the Earth's surface, arose in a wrinkling of the crust under the influence of cooling and skrinkage of the subcrustal materials, is held by many eminent geologists, but not without dissent from others.

The most striking observational fact connected with mountain structure is that, without exception, the ranges

of the Earth are built essentially of sedimentary rocks: that is of rocks which have been accumulated at some remote past time beneath the surface of the ocean. A volcanic core there may sometimes be—probably an attendant or consequence of the uplifting—or a core of plutonic igneous rocks which has arisen under the same compressive forces which have bowed and arched the strata from their original horizontal position. It is not uncommon to meet among unobservant people those who regard all mountain ranges as volcanic in origin. Volcanoes, however, do not build mountain ranges. They break out as more or less isolated cones or hills. Compare the map of the Auvergne with that of Switzerland; the volcanoes of South Italy with the Apennines. Such great ranges as those which border with triple walls the west coast of North America are in no sense volcanic: nor are the Pyrenees, the Caucasus, or the Himalaya. Volcanic materials are poured out from the summits of the Andes, but the range itself is built up of folded sediments on the same architecture as the other great ranges of the Earth.

Before attempting an explanation of the origin of the mountains we must first become more closely acquainted with the phenomena attending mountain elevation.

At the present day great accumulations of sediment are taking place along the margins of the continents where the rivers reach the ocean. Thus, the Gulf of Mexico receiving the sediment of the Mississippi and Rio Grande;

the northeast coast of South America receiving the sediments of the Amazons; the east coast of Asia receiving the detritus of the Chinese rivers; are instances of such areas of deposition. Year by year, century by century, the accumulation progresses, and as it grows the floor of the sea sinks under the load. Of the yielding of the crust under the burthen of the sediments we are assured; for otherwise the many miles of vertically piled strata which are uplifted to our view in the mountains, never could have been deposited in the coastal seas of the past. The flexure and sinking of the crust are undeniable realities.

Such vast subsiding areas are known as geosynclines. From the accumulated sediments of the geosynclines the mountain ranges of the past have in every case originated; and the mountains of the future will assuredly arise and lofty ranges will stand where now the ocean waters close over the collecting sediments. Every mountain range upon the Earth enforces the certainty of this prediction.

The mountain-forming movement takes place after a certain great depth of sediment is collected. It is most intense where the thickness of deposit is greatest. We see this when we examine the structure of our existing mountain ranges. At either side where the sediments thin out, the disturbance dies away, till we find the comparatively shallow and undisturbed level sediments which clothe the continental surface.

Whatever be the connection between the deposition and

the subsequent upheaval, _the element of great depth of accumulation seems a necessary condition and must evidently enter as a factor into the Physical Processes involved_. The mountain range can only arise where the geosyncline is deeply filled by long ages of sedimentation.

Dana's description of the events attending mountain building is impressive:

"A mountain range of the common type, like that to which the Appalachians belong, is made out of the sedimentary formations of a long preceding era; beds that were laid down conformably, and in succession, until they had reached the needed thickness; beds spreading over a region tens of thousands of square miles in area. The region over which sedimentary formations were in progress in order to make, finally, the Appalachian range, reached from New York to Alabama, and had a breadth of 100 to 200 miles, and the pile of horizontal beds along the middle was 40,000 feet in depth. The pile for the Wahsatch Mountains was 60,000 feet thick, according to King. The beds for the Appalachians were not laid down in a deep ocean, but in shallow waters, where a gradual subsidence was in progress; and they at last, when ready for the genesis, lay in a trough 40,000 feet deep, filling the trough to the brim. It thus appears that epochs of mountain-making have occurred only after long intervals of quiet in the history of a continent."[1]

[1] Dana, _Manual of Geology_, third edition, p. 794

On the western side of North America the work of mountain-building was, indeed, on the grandest scale. For long ages and through a succession of geological epochs, sedimentation had proceeded so that the accumulations of Palaeozoic and Mesozoic times had collected in the geosyncline formed by their own ever increasing weight. The site of the future Laramide range was in late Cretaceous times occupied by some 50,000 feet of sedimentary deposits; but the limit had apparently been attained, and at this time the Laramide range, as well as its southerly continuation into the United States, the Rockies, had their beginning. Chamberlin and Salisbury[1] estimate that the height of the mountains developed in the Laramide range at this time was 20,000 feet, and that, owing to the further elevation which has since taken place, from 32,000 to 35,000 feet would be their present height if erosion had not reduced them. Thus on either side of the American continent we have the same forces at work, throwing up mountain ridges where the sediments had formerly been shed into the ocean.

These great events are of a rhythmic character; the crust, as it were, pulsating under the combined influences of sedimentation and denudation. The first involves downward movements under a gathering load, and ultimately a reversal of the movement to one of upheaval; the second factor, which throughout has been in

[1] Chamberlin and Salisbury, _Geology_, 1906, iii., 163.

operation as creator of the sediments, then intervenes as an assailant of the newly-raised mountains, transporting their materials again to the ocean, when the rhythmic action is restored to its first phase, and the age-long sequence of events must begin all over again.

It has long been inferred that compressive stress in the crust must be a primary condition of these movements. The wvork required to effect the upheavals must be derived from some preexisting source of energy. The phenomenon—intrinsically one of folding of the crust—suggests the adjustment of the earth-crust to a lessening radius; the fact that great mountain-building movements have simultaneously affected the entire earth is certainly in favour of the view that a generally prevailing cause is at the basis of the phenomenon.

The compressive stresses must be confined to the upper few miles of the crust, for, in fact, the downward increase of temperature and pressure soon confers fluid properties on the medium, and slow tangential compression results in hydrostatic pressure rather than directed stresses. Thus the folding visible in the mountain range, and the lateral compression arising therefrom, are effects confined to the upper parts of the crust.

The energy which uplifts the mountain is probably a surviving part of the original gravitational potential energy of the crust itself. It must be assumed that the crust in following downwards the shrinking subcrustal magma, develops immense compressive stresses in

its materials, vast geographical areas being involved. When folding at length takes place along the axis of the elongated syncline of deposition, the stresses find relief probably for some hundreds of miles, and the region of folding now becomes compressed in a transverse direction. As an illustration, the Laramide range, according to Dawson, represents the reduction of a surface-belt 50 miles wide to one of 25 miles. The marvellous translatory movements of crustal folds from south to north arising in the genesis of the Swiss Alps, which recent research has brought to light, is another example of these movements of relief, which continue to take place perhaps for many millions of years after they are initiated.

The result of this yielding of the crust is a buckling of the surface which on the whole is directed upwards; but depression also is an attendant, in many cases at least, on mountain upheaval. Thus we find that the ocean floor is depressed into a syncline along the western coast of South America; a trough always parallel to the ranges of the Andes. The downward deflection of the crust is of course an outcome of the same compressive stresses which elevate the mountain.

The fact that the yielding of the crust is always situated where the sediments have accumulated to the greatest depth, has led to attempts from time to time of establishing a physical connexion between the one and the other. The best-known of these theories is that of Babbage and Herschel. This seeks the connexion in the rise of the geotherms into the sinking mass of sediment and the consequent increase of temperature of the earth-crust beneath. It will be understood that as these isogeotherms, or levels at which the temperature is the same, lie at a uniform distance from the surface all over the Earth, unless where special variations of conductivity may disturb them, the introduction of material pressed downwards from above must result in these materials partaking of the temperature proper to the depth to which they are depressed. In other words the geotherms rise into the sinking sediments, always, however, preserving their former average distance from the surface. The argument is that as this process undoubtedly involves the heating up of that portion of the crust which the sediments have displaced downwards, the result must be a local enfeeblement of the crust, and hence these areas become those of least resistance to the stresses in the crust.

When this theory is examined closely, we see that it only amounts to saying that the bedded rocks, which have taken the place of the igneous materials beneath, as a part of the rigid crust of the Earth, must be less able to withstand compressive stress than the average crust. For there has been no absolute rise of the geotherms, the thermal conductivities of both classes of materials differing but little. Sedimentary rock has merely taken the place of average crust-rock, and is subjected to the same average temperature and pressure prevailing in the surrounding crust. But are there any grounds for the

assumption that the compressive resistance of a complex of sedimentary rocks is inferior to one of igneous materials? The metamorphosed siliceous sediments are among the strongest rocks known as regards resistance to compressive stress; and if limestones have indeed plastic qualities, it must be remembered that their average amount is only some 5 per cent. of the whole. Again, so far as rise of temperature in the upper crust may affect the question, a temperature which will soften an average igneous rock will not soften a sedimentary rock, for the reason that the effect of solvent denudation has been to remove those alkaline silicates which confer fusibility.

When, however, we take into account the radioactive content of the sediments the matter assumes a different aspect.

The facts as to the general distribution of radioactive substances at the surface, and in rocks which have come from considerable depths in the crust, lead us to regard as certain the widespread existence of heat-producing radioactive elements in the exterior crust of the Earth. We find, indeed, in this fact an explanation—at least in part—of the outflow of heat continually taking place at the surface as revealed by the rising temperature inwards. And we conclude that there must be a thickness of crust amounting to some miles, containing the radioactive elements.

Some of the most recent measurements of the quantities of radium and thorium in the rocks of igneous origin—_e.g._ granites, syenites, diorites, basalts, etc., show that the

radioactive heat continually given out by such rocks amounts to about one millionth part of 0.6 calories per second per cubic metre of average igneous rock. As we have to account for the escape of about 0.0014 calorie[1] per square metre of the Earth's surface per second (assuming the rise of temperature downwards, _i.e._ the "gradient" of temperature, to be one degree centigrade in 35 metres) the downward extension of such rocks might, _prima facie_, be as much as 19 kilometres.

About this calculation we have to observe that we assume the average radioactivity of the materials with which we have dealt at the surface to extend uniformly all the way down, _i.e._ that our experiments reveal the average radioactivity of a radioactive crust. There is much to be said for this assumption. The rocks which enter into the measurements come from all depths of the crust. It is highly probable that the less silicious, _i.e._ the more basic, rocks, mainly come from considerable depths; the more acid or silica-rich rocks, from higher levels in the crust. The radioactivity determined as the mean of the values for these two classes of rock closely agrees with that found for intermediate rocks, or rocks containing an intermediate amount of silica. Clarke contends that this last class of material probably represents the average composition of the Earth's crust so far as it has been explored by us.

[1] The calorie referred to is the quantity of heat required to heat one gram of water, _i.e._ one cubic centimetre of water—through one degree centigrade.

It is therefore highly probable that the value found for the mean radioactivity of acid and basic rocks, or that found for intermediate rocks, truly represents the radioactive state of the crust to a considerable depth. But it is easy to show that we cannot with confidence speak of the thickness of this crust as determinable by equating the heat outflow at the surface with the heat production of this average rock.

This appears in the failure of a radioactive layer, taken at a thickness of about 19-kilometres, to account for the deep-seated high temperatures which we find to be indicated by volcanic phenomena at many places on the surface. It is not hard to show that the 19-kilometre layer would account for a temperature no higher than about 270° >C. at its base.

It is true that this will be augmented beneath the sedimentary deposits as we shall presently see; and that it is just in association with these deposits that deep-seated temperatures are most in evidence at the surface; but still the result that the maximum temperature beneath the crust in general attains a value no higher than 270° C. is hardly tenable. We conclude, then, that some other source of heat exists beneath. This may be radioactive in origin and may be easily accounted for if the radioactive materials are more sparsely distributed at the base of the upper crust. Or, again, the heat may be primeval or original heat, still escaping from a cooling world. For our present purpose it does not much matter which view

we adopt. But we must recognise that the calculated depth of 19 kilometres of crust, possessing the average radioactivity of the surface, is excessive; for, in fact, we are compelled by the facts to recognise that some other source of heat exists beneath.

If the observed surface gradient of temperature persisted uniformly downwards, at some 35 kilometres beneath the surface there would exist temperatures (of about 1000° C.) adequate to soften basic rocks. It is probable, however, that the gradient diminishes downwards, and that the level at which such temperatures exist lies rather deeper than this. It is, doubtless, somewhat variable according to local conditions; nor can we at all approximate closely to an estimate of the depth at which the fusion temperatures will be reached, for, in fact, the existence of the radioactive layer very much complicates our estimates. In what follows we assume the depth of softening to lie at about 40 kilometres beneath the surface of the normal crust; that is 25 miles down. It is to be observed that Prestwich and other eminent geologists, from a study of the facts of crust-folding, etc., have arrived at similar estimates.[1] As a further assumption we are probably not far wrong if we assign to the radioactive part of this crust a thickness of about 10 or 12 kilometres; _i.e._ six or seven miles. This is necessarily a rough approximation only; but the conclusions at which

[1] Prestwich, _Proc. Royal Soc._, xii., p. 158 _et seq._

we shall arrive are reached in their essential features allowing a wide latitude in our choice of data. We shall speak of this part of the crust as the normal radioactive layer.

An important fact is evolved from the mathematical investigation of the temperature conditions arising from the presence of such a radioactive layer. It is found that the greatest temperature, due to the radioactive heat everywhere evolved in the layer—_i.e._ the temperature at its base—is proportional to the square of the thickness of the layer. This fact has a direct bearing on the influence of radioactivity upon mountain elevation; as we shall now find.

The normal radioactive layer of the Earth is composed of rocks extending—as we assume—approximately to a depth of 12 kilometres (7.5 miles). The temperature at the base of this layer due to the heat being continually evolved in it, is, say, t_1°. Now, let us suppose, in the trough of the geosyncline, and upon the top of the normal layer, a deposit of, say, 10 kilometres (6.2 miles) of sediments is formed during a long period of continental denudation. What is the effect of this on the temperature at the base of the normal layer depressed beneath this load? The total thickness of radioactive rocks is now 22 kilometres. Accordingly we find the new temperature t_2°, by the proportion t_1° : t_2° :: 12° : 22° That is, as 144 to 484. In fact, the temperature is more than trebled. It is true we here assume the radioactivity of the sediments

and of the normal crust to be the same. The sediments are, however, less radioactive in the proportion of 4 to 3. Nevertheless the effects of the increased thickness will be considerable.

Now this remarkable increase in the temperature arises entirely from the condition attending the radioactive heating; and involves something _additional_ to the temperature conditions determined by the mere depression and thickening of the crust as in the Babbage-Herschel theory. The latter theory only involves a _shifting_ of the temperature levels (or geotherms) into the deposited materials. The radioactive theory involves an actual rise in the temperature at any distance from the surface; so that _the level in the crust at which the rocks are softened is nearer to the surface in the geosynclines than it is elsewhere in the normal crust_ (Pl. XV, p. 118).

In this manner the rigid part of the crust is reduced in thickness where the great sedimentary deposits have collected. A ten-kilometre layer of sediment might result in reducing the effective thickness of the crust by 30 per cent.; a fourteen-kilometre layer might reduce it by nearly 50 per cent. Even a four-kilometre deposit might reduce the effective resistance of the crust to compressive forces, by 10 per cent.

Such results are, of course, approximate only. They show that as the sediments grow in depth there is a rising of the geotherm of plasticity—whatever its true temperature may be—gradually reducing the thickness of that part

of the upper crust which is bearing the simultaneously increasing compressive stresses. Below this geotherm long-continued stress resolves itself into hydrostatic pressure; above it (there is, of course, no sharp line of demarcation) the crust accumulates elastic energy. The final yielding and flexure occur when the resistant cross-section has been sufficiently diminished. It is probable that there is also some outward hydrostaitic thrust over the area of rising temperature, which assists in determining the upward throw of the folds.

When yielding has begun in any geosyncline, and the materials are faulted and overthrust, there results a considerably increased thickness. As an instance, consider the piling up of sediments over the existing materials of the Alps, which resulted from the compressive force acting from south to north in the progress of Alpine upheaval. Schmidt of Basel has estimated that from 15 to 20 kilometres of rock covered the materials of the Simplon as now exposed, at the time when the orogenic forces were actively at work folding and shearing the beds, and injecting into their folds the plastic gneisses coming from beneath.[1] The lateral compression of the area of deposition of the Laramide, already referred to, resulted in a great thickening of the deposits. Many other cases might be cited; the effect is always in some degree necessarily produced.

[1] Schmidt, Ec. Geol. _Helvelix_, vol. ix., No. 4, p. 590

If time be given for the heat to accumulate in the lower depths of the crushed-up sediments, here is an additional source of increased temperature. The piled-up masses of the Simplon might have occasioned a rise due to radioactive heating of one or two hundred degrees, or even more; and if this be added to the interior heat, a total of from 800° to 1000° might have prevailed in the rocks now exposed at the surface of the mountain. Even a lesser temperature, accompanied by the intense pressure conditions, might well occasion the appearances of thermal metamorphism described by Weinschenk, and for which, otherwise, there is difficulty in accounting.[1]

This increase upon the primarily developed temperature conditions takes place concurrently with the progress of compression; and while it cannot be taken into account in estimating the conditions of initial yielding of the crust, it adds an element of instability, inasmuch as any progressive thickening by lateral compression results in an accelerated rise of the goetherms. It is probable that time sufficient for these effects to develop, if not to their final, yet to a considerable extent, is often available. The viscous movements of siliceous materials, and the out-pouring of igneous rocks which often attend mountain elevation, would find an explanation in such temperatures.

[1] Weinschenk, _Congrès Géol. Internat._, 1900, i., p. 332.

There is no more striking feature of the part here played by radioactivity than the fact that the rhythmic occurrence of depression and upheaval succeeding each other after great intervals of time, and often shifting their position but little from the first scene of sedimentation, becomes accounted for. The source of thermal energy, as we have already remarked, is in fact transported with the sediments—that energy which determines the place of yielding and upheaval, and ordains that the mountain ranges shall stand around the continental borders. Sedimentation from this point of view is a convection of energy.

When the consolidated sediments are by these and by succeeding movements forced upwards into mountains, they are exposed to denudative effects greatly exceeding those which affect the plains. Witness the removal during late Tertiary times of the vast thickness of rock enveloping the Alps. Such great masses are hurried away by ice, rivers, and rain. The ocean receives them; and with infinite patience the world awaits the slow accumulation of the radioactive energy beginning afresh upon its work. The time for such events appears to us immense, for millions of years are required for the sediments to grow in thickness, and the geotherms to move upwards; but vast as it is, it is but a moment in the life of the parent radioactive substances, whose atoms, hardly diminished in numbers, pursue their changes while the mountains come and go, and the

rudiments of life develop into its highest consummations.

To those unacquainted with the results of geological investigation the history of the mountains as deciphered in the rocks seems almost incredible.

The recently published sections of the Himalaya, due to H. H. Hayden and the many distinguished men who have contributed to the Geological Survey of India, show these great ranges to be essentially formed of folded sediments penetrated by vast masses of granite and other eruptives. Their geological history may be summarised as follows

The Himalayan area in pre-Cambrian times was, in its southwestern extension, part of the floor of a sea which covered much of what is now the Indian Peninsula. In the northern shallows of this sea were laid down beds of conglomerate, shale, sandstone and limestone, derived from the denudation of Archæan rocks, which, probably, rose as hills or mountains in parts of Peninsular India and along the Tibetan edge of the Himalayan region. These beds constitute the record of the long Purana Era[1] and are probably coeval with the Algonkian of North America. Even in these early times volcanic disturbances affected this area and the lower beds of the Purana deposits were penetrated by volcanic outflows, covered by sheets of lava, uplifted, denuded and again submerged

[1] See footnote, p. 139.

beneath the waters. Two such periods of instability have left their records in the sediments of the Purana sea.

The succeeding era—the Dravidian Era—opens with Haimanta (Cambrian) times. A shallow sea now extended over Kumaun, Garwal, and Spiti, as well as Kashmir and ultimately over the Salt Range region of the Punjab as is shown by deposits in these areas. This sea was not, however, connected with the Cambrian sea of Europe. The fossil faunas left by the two seas are distinct.

After an interval of disturbance during closing Haimanta times, geographical changes attendant on further land movements occurred. The central sea of Asia, the Tethys, extended westwards and now joined with the European Paleozoic sea; and deposits of Ordovician and Silurian age were laid down:—the Muth deposits.

The succeeding Devonian Period saw the whole Northern Himalayan area under the waters of the Tethys which, eastward, extended to Burma and China and, westward, covered Kashmir, the Hindu Kush and part of Afghanistan. Deposits continued to be formed in this area till middle Carboniferous times.

Near. the close of the Dravidian Era Kashmir became convulsed by volcanic disturbance and the Penjal traps were ejected. It was a time of worldwide disturbance and of redistribution of land and water. Carboniferous times had begun, and the geographical changes in

the southern limits of the Tethys are regarded as ushering in a new and last era in Indian geological history the Aryan Bra.

India was now part of Gondwanaland; that vanished continent which then reached westward to South Africa and eastward to Australia. A boulder-bed of glacial origin, the Talchir Boulder-bed, occurs in many surviving parts of this great land. It enters largely into the Salt Range deposits. There is evidence that extensive sheets of ice, wearing down the rocks of Rajputana, shoved their moraines northward into the Salt Range Sea; then, probably, a southern extension of the Tethys.

Subsequent to this ice age the Indian coalfields of the Gondwana were laid down, with beds rich in the Glossopteris and Gangamopteris flora. This remarkable carboniferous flora extends to Southern Kashmir, so that it is to be inferred that this region was also part of the main Gondwanaland. But its emergence was but for a brief period. Upper Carboniferous marine deposits succeeded; and, in fact, there was no important discontinuity in the deposits in this area from Panjal times till the early Tertiaries. During the whole of which vast period Kashmir was covered with the waters of the Tethys.

The closing Dravidian disturbances of the Kashmir region did not, apparently, extend to the eastern Himalayan area. But the Carboniferous Period was, in this

eastern area, one of instability, culminating, at the close of the Period, in a steady rise of the land and a northward retreat of the Tethys. Nearly the entire Himalaya east of Kashmir became a land surface and remained exposed to denudative forces for so long a time that in places the whole of the Carboniferous, Devonian, and a large part of the Silurian and Ordovician deposits were removed—some thousands of feet in thickness—before resubmergence in the Tethys occurred.

Towards the end of the Palaeozoic Age the Aryan Tethys receded westwards, but still covered the Himalaya and was still connected with the European Palæozoic sea. The Himalayan area (as well as Kashmir) remained submerged in its waters throughout the entire Mesozoic Age.

During Cretaceous times the Tethys became greatly extended, indicating a considerable subsidence of northwestern India, Afghanistan, Western Asia, and, probably, much of Tibet. The shallow-water character of the deposits of the Tibetan Himalaya indicates, however, a coast line near this region. Volcanic materials, now poured out, foreshadow the incoming of the great mountain-building epoch of the Tertiary Era. The enormous mass of the Deccan traps, possessing a volume which has been estimated at as much as 6,000 cubic miles, was probably extruded over the Northern Peninsular region during late Cretaceous times. The sea now began to retreat, and by the close of

the Eocene, it had moved westward to Sind and Baluchistan. The movements of the Earth's crust were attended by intense volcanic activity, and great volumes of granite were injected into the sediments, followed by dykes and outflows of basic lavas.

The Tethys vanished to return no more. It survives in the Mediterranean of today. The mountain-building movements continued into Pliocene times. The Nummulite beds of the Eocene were, as the result, ultimately uplifted 18,500 feet over sea level, a total uplift of not less than 20,000 feet.

Thus with many vicissitudes, involving intervals of volcanic activity, local uplifting, and extensive local denudation, the Himalaya, which had originated in the sediments of the ancient Purana sea, far back in pre-Cambrian times, and which had developed potentially in a long sequence of deposits collecting almost continuously throughout the whole of geological time, finally took their place high in the heavens, where only the winds—faint at such altitudes—and the lights of heaven can visit their eternal snows.[1]

In this great history it is significant that the longest continuous series of sedimentary deposits which the world has known has become transfigured into the loftiest elevation upon its surface.

[1] See A Sketch of the _Geography and Geology of the Himalaya Mountains and Tibet_. By Colonel S. G. Burrard, R.E., F.R.S., and H. H. Hayden, F.G.S., Part IV. Calcutta, 1908.

The diagrammatic sections of the Himalaya accompanying this brief description arc taken from the monograph of Burrard and Hayden (loc. cit.) on the Himalaya. Looking at the sections we see that some of the loftiest summits are sculptured in granite and other crystalline rocks. The appearance of these materials at the surface indicates the removal by denudation and the extreme metamorphism of much sedimentary deposit. The crystalline rocks, indeed, penetrate some of the oldest rocks in the world. They appear in contact with Archaean, Algonkian or early Palaeozoic rocks. A study of the sections reveals not only the severe earth movements, but also the immense amount of sedimentary deposits involved in the genesis of these alps. It will be noted that the vertical scale is not exaggerated relatively to the horizontal.[1] Although there is no evidence of mountain building

[1] To those unacquainted with the terminology of Indian geology the following list of approximate equivalents in time will be of use

Ngari Khorsum Beds - Pleistocene.
Siwalik Series - Miocene and Pliocene.
Sirmur Series - Oligocene.
Kampa System - Eocene and Cretaceous.
Lilang System - Triassic.
Kuling System - Permian.
Gondwana System - Carboniferous.
Kenawar System - Carboniferous and Devonian
Muth System - Silurian.
Haimanta System - Mid. and Lower Cambrian.
Purana Group - Algonkian.
Vaikrita System - Archæan.
Daling Series - Archæan.

on a large scale in the Himalayan area till the Tertiary upheaval, it is, in the majority of cases, literally correct to speak of the mountains as having their generations like organic beings, and passing through all the stages of birth, life, death and reproduction. The Alps, the Jura, the Pyrenees, the Andes, have been remade more than once in the course of geological time, the _débris_ of a worn-out range being again uplifted in succeeding ages.

Thus to dwell for a moment on one case only: that of the Pyrenees. The Pyrenees arose as a range of older Palmozoic rocks in Devonian times. These early mountains, however, were sufficiently worn out and depressed by Carboniferous times to receive the deposits of that age laid down on the up-turned edges of the older rocks. And to Carboniferous succeeded Permian, Triassic, Jurassic and Lower Cretaceous sediments all laid down in conformable sequence. There was then fresh disturbance and upheaval followed by denudation, and these mountains, in turn, became worn out and depressed beneath the ocean so that Upper Greensand rocks were laid down unconforrnably on all beneath. To these now succeeded Upper Chalk, sediments of Danian age, and so on, till Eocene times, when the tale was completed and the existing ranges rose from the sea. Today we find the folded Nummulitic strata of Eocene age uplifted 11,000 feet, or within 200 feet of the greatest heights of the Pyrenees. And so they stand awaiting

the time when once again they shall "fall into the portion of outworn faces."[1]

Only mountains can beget mountains. Great accumulations of sediment are a necessary condition for the localisation of crust-flexure. The earliest mountains arose as purely igneous or volcanic elevations, but the generations of the hills soon originated in the collection of the _débris_, under the law of gravity, in the hollow places. And if a foundered range is exposed now to our view encumbered with thousands of feet of overlying sediments we know that while the one range was sinking, another, from which the sediments were derived, surely existed. Through the "windows" in the deep-cut rocks of the Swiss valleys we see the older Carboniferous Alps looking out, revisiting the sun light, after scores of millions of years of imprisonment. We know that just as surely as the Alps of today are founding by their muddy torrents ranges yet to arise, so other primeval Alps fed into the ocean the materials of these buried pre-Permian rocks.

This succession of events only can cease when the rocks have been sufficiently impoverished of the heat-producing substances, or the forces of compression shall have died out in the surface crust of the earth.

It seems impossible to escape the conclusion that in the great development of ocean-encircling areas of

[1] See Prestwich, _Chemical and Physical Geology_, p. 302.

deposition and crustal folding, the heat of radioactivity has been a determining factor. We recognise in the movements of the sediments not only an influence localising and accelerating crustal movements, but one which, in subservience to the primal distribution of land and water, has determined some of the greatest geographical features of the globe.

It is no more than a step to show that bound up with the radioactive energy are most of the earthquake and volcanic phenomena of the earth. The association of earthquakes with the great geosynclines is well known. The work of De Montessus showed that over 94 per cent. of all recorded shocks lie in the geosynclinal belts. There can be no doubt that these manifestations of instability are the results of the local weakness and flexure which originated in the accumulation of energy denuded from the continents. Similarly we may view in volcanoes phenomena referable to the same fundamental cause. The volcano was, in fact, long regarded as more intimately connected with earthquakes than it, probably, actually is; the association being regarded in a causative light, whereas the connexion is more that of possessing a common origin. The girdle of volcanoes around the Pacific and the earthquake belt coincide. Again, the ancient and modern volcanoes and earthquakes of Europe are associated with the geosyncline of the greater Mediterranean, the Tethys of Mesozoic times. There is no difficulty in understanding in a

general way the nature of the association. The earthquake is the manifestation of rupture and slip, and, as Suess has shown, the epicentres shift along that fault line where the crust has yielded.[1] The volcano marks the spot where the zone of fusion is brought so high in the fractured crust that the melted materials are poured out upon the surface.

In a recent work on the subject of earthquakes Professor Hobbs writes: "One of the most interesting of the generalisations which De Montessus has reached as a result of his protracted studies, is that the earthquake districts on the land correspond almost exactly to those belts upon the globe which were the almost continuous ocean basins of the long Secondary era of geological history. Within these belts the sedimentary formations of the crust were laid down in the greatest thickness, and the formations follow each other in relatively complete succession. For almost or quite the whole of this long era it is therefore clear that the ocean covered these zones. About them the formations are found interrupted, and the lacuna indicate that the sea invaded the area only to recede from it, and again at some later period to transgress upon it. For a long time, therefore, these earthquake belts were the sea basins—the geosynclines. They became later the rising mountains of the Tertiary period, and mountains they

[1] Suess, _The Face of the Earth_, vol. ii., chap. ii.

are today. The earthquake belts are hence those portions of the earth's crust which in recent times have suffered the greatest movements in a vertical direction—they are the most mobile portions of the earth's crust."[1] Whether the movements attending mountain elevation and denudation are a connected and integral part of those wide geographical changes which result in submergence and elevation of large continental areas, is an obscure and complex question. We seem, indeed, according to the views of some authorities, hardly in a position to affirm with certainty that such widespread movements of the land have actually occurred, and that the phenomena are not the outcome of fluctuations of oceanic level; that our observations go no further than the recognition of positive and negative movements of the strand. However this may be, the greater part of mechanical denudation during geological time has been done on the mountain ranges. It is, in short, indisputable that the orogenic movements which uplift the hills have been at the basis of geological history. To them the great accumulations of sediments which now form so large a part of continental land are mainly due. There can be no doubt of the fact that these movements have swayed the entire history, both inorganic and organic, of the world in which we live.

[1] Hobbs, _Earthquakes_, p. 58.

To sum the contents of this essay in the most general terms, we find that in the conception of denudation as producing the convection and accumulation of radiothermal energy the surface features of the globe receive a new significance. The heat of the earth is not internal only, but rather a heat-source exists at the surface, which, as we have seen, cannot prevail to the same degree within; and when the conditions become favourable for the aggregation of the energy, the crust, heated both from beneath and from above, assumes properties more akin to those of its earlier stages of development, the secular heat-loss being restored in the radioactive supplies. These causes of local mobility have been in operation, shifting somewhat from place to place, and defined geographically by the continental masses undergoing denudation, since the earliest times.

ALPINE STRUCTURE

AN intelligent observer of the geological changes progressing in southern Europe in Eocene times would have seen little to inspire him with a premonition of the events then developing. The Nummulitic limestones were being laid down in that enlarged Mediterranean which at this period, save for a few islands, covered most of south Europe. Of these stratified remains, as well as of the great beds of Cretaceous, Jurassic, Triassic, and Permian sediments beneath, our hypothetical observer would probably have been regardless; just as today we observe, with an indifference born of our transitoriness, the deposits rapidly gathering wherever river discharge is distributing the sediments over the sea-floor, or the lime-secreting organisms are actively at work. And yet it took but a few millions of years to uplift the deposits of the ancient Tethys; pile high its sediments in fold upon fold in the Alps, the Carpathians, and the Himalayas; and—exposing them to the rigours of denudation at altitudes where glaciation, landslip, and torrent prevail—inaugurate a new epoch of sedimentation and upheaval.

In the case of the Alps, to which we wish now specially to refer, the chief upheaval appears to have been in Oligocene times, although movement continued to the close of the Pliocene. There was thus a period of some millions of years within which the entire phenomena were comprised. Availing ourselves of Sollas' computations,[1] we may sum the maximum depths of sedimentary deposits of the geological periods concerned as follows:—

Pliocene - - - - - 3,950 m.

Miocene - - - - - 4,250 m.

Oligocene - - - - 3,660 m.

Eocene - - - - - - 6,100 m.

and assuming that the orogenic forces began their work in the last quarter of the Eocene period, we have a total of 13,400 m. as some measure of the time which elapsed. At the rate of io centimetres in a century these deposits could not have collected in less than 13.4 millions of years. It would appear that not less than some ten millions of years were consumed in the genesis of the Alps before constructive movements finally ceased.

The progress of the earth-movements was attended by the usual volcanic phenomena. The Oligocene and Miocene volcanoes extended in a band marked by the Auvergne, the Eiffel, the Bohemian, and the eastern Carpathian eruptions; and, later, towards the close of the movements in Pliocene times, the south border

[1] Sollas, Anniversary Address, Geol. Soc., London, 1909.

regions of the Alps became the scene of eruptions such as those of Etna, Santorin, Somma (Vesuvius), etc.

We have referred to these well-known episodes with two objects in view: to recall to mind the time-interval involved, and the evidence of intense crustal disturbance, both dynamic and thermal. According to views explained in a previous essay, the energetic effects of radium in the sediments and upper crust were a principal factor in localising and bringing about these results. We propose now to inquire if, also, in the more intimate structure of the Alps, the radioactive energy may not have borne a part.

What we see today in the Alps is but a residue spared by denudation. It is certain that vast thicknesses of material have disappeared. Even while constructive effects were still in progress, denudative forces were not idle. Of this fact the shingle accumulations of the Molasse, where, on the northern borders of the Alps, they stand piled into mountains, bear eloquent testimony. In the sub-Apennine series of Italy, the great beds of clays, marls, and limestones afford evidence of these destructive processes continued into Pliocene times. We have already referred to Schmidt's estimate that the sedimentary covering must have in places amounted to from 15,000 to 20,000 metres. The evidence for this is mainly tectonic or structural; but is partly forthcoming in the changes which the materials now open to our inspection plainly reveal. Thus it is impos-

sible to suppose that gneissic rocks can become so far plastic as to flow in and around the calcareous sediments, or be penetrated by the latter—as we see in the Jungfrau and elsewhere—unless great pressures and high temperatures prevailed. And, according to some writers, the temperatures revealed by the intimate structural changes of rock-forming minerals must have amounted to those of fusion. The existence of such conditions is supported by the observation that where the.crystallisation is now the most perfect, the phenomena of folding and injection are best developed.[1] These high temperatures would appear to be unaccountable without the intervention of radiothermal effects; and, indeed, have been regarded as enigmatic by observers of the phenomena in question. A covering of 20,000 metres in thickness would not occasion an earth-temperature exceeding 500° C. if the gradients were such as obtain in mountain regions generally; and 600° is about the limit we could ascribe to the purely passive effects of such a layer in elevating the geotherms.

Those who are still unacquainted with the recently published observations on the structure of the Alps may find it difficult to enter into what has now to be stated; for the facts are, indeed, very different from the generally preconceived ideas of mountain formation. Nor can we wonder that many geologists for long held

[1] Weinschenk, C. R. _Congrès Géol._, 1900, p. 321, et seq.

back from admitting views which appeared so extreme. Receptivity is the first virtue of the scientific mind; but, with every desire to lay aside prejudice, many felt unequal to the acceptance of structural features involving a folding of the earth-crust in laps which lay for scores of miles from country to country, and the carriage of mountainous materials from the south of the Alps to the north, leaving them finally as Alpine ranges of ancient sediments reposing on foundations of more recent date. The historian of the subject will have to relate how some who finally were most active in advancing the new views were at first opposed to them. In the change of conviction of these eminent geologists we have the strongest proof of the convincing nature of the observations and the reality of the tectonic features upon which the recent views are founded.

The lesser mountains which stand along the northern border of the great limestone Alps, those known as the Préalpes, present the strange characteristic of resting upon materials younger than themselves. Such mountains as the remarkable-looking Mythen, near Schwyz, for instance, are weathered from masses of Triassic and

Jurassic rock, and repose on the much more recent Flysch. In sharp contrast to the Flysch scenery, they stand as abrupt and gigantic erratics, which have been transported from the central zone of the Alps lying far to the south. They are strangers petrologically,

stratigraphically, and geographically,[1] to the locality in which they now occur. The exotic materials may be dolomites, limestones, schists, sandstones, or rocks of igneous origin. They show in every case traces of the severe dynamic actions to which they have been subjected in transit. The igneous, like the sedimentary, klippen, can be traced to distant sources; to the massif of Belladonne, to Mont Blanc, Lugano, and the Tyrol. The Préalpes are, in fact, mountains without local roots.

In this last-named essential feature, the Préalpes do not differ from the still greater limestone Alps which succeed them to the south. These giants, _e.g._ the Jungfrau, Wetterhorn, Eiger, etc., are also without local foundations. They have been formed from the overthrown and drawn-out anticlines of great crust-folds, whose synclines or roots are traceable to the south side of the Rhone Valley. The Bernese Oberland originated in the piling-up of four great sheets or recumbent folds, one of which is continued into the Préalpes. With Lugeon[2] we may see in the phenomenon of the formation of the Préalpes a detail; regarding it as a normal expression of that mechanism which has created the Swiss Alps. For these limestone masses of the Oberland are not indications of a merely local shift of the sedimentary covering of the Alps. Almost the whole covering has

[1] De Lapparent, _Traité de Géologie_, p. 1,785.

[2] Lugeon, _Bulletin Soc. Géol. de France_, 1901, p. 772.

been pushed over and piled up to the north. Lugeon[l] concludes that, before denudation had done its work and cut off the Préalpes from their roots, there would have been found sheets, to the number of eight, superimposed and extending between the Mont Blanc massif and the massif of the Finsteraarhorn: these sheets being the overthrown folds of the wrinkled sedimentary covering. The general nature of the alpine structure

{Fig. 8}

will be understood from the presentation of it diagrammatically after Schmidt of Basel (Fig. 8).[2] The section extends from north to south, and brings out the relations of the several recumbent folds. We must imagine almost the whole of these superimposed folds now removed from the central regions of the Alps by denudation,

[1] Lugeon, _loc. cit._

[2] Schmidt, _Ec. Geol. Helvetiae_, vol. ix., No. 4.

and leaving the underlying gneisses rising through the remains of Permian, Triassic, and Jurassic sediments; while to the north the great limestone mountains and further north still, the Préalpes, carved from the remains of the recumbent folds, now stand with almost as little resemblance to the vanished mountains as the memories of the past have to its former intense reality.

These views as to the origin of the Alps, which are shared at the present day by so many distinguished geologists, had their origin in the labours of many now gone; dating back to Studer; finding their inspiration in the work of Heim, Suess, and Marcel Bertrand; and their consummation in that of Lugeon, Schardt, Rothpletz, Schmidt, and many others. Nor must it be forgotten that nearer home, somewhat similar phenomena, necessarily on a smaller scale, were recognised by Lapworth, twenty-six years ago, in his work on the structure of the Scottish Highlands.

An important tectonic principle underlies the development of the phenomena we have just been reviewing. The uppermost of the superimposed recumbent folds is more extended in its development than those which lie beneath. Passing downwards from the highest of the folds, they are found to be less and less extended both in the northerly and in the southerly direction, speaking of the special case—the Alps—now before us. This feature might be described somewhat differently. We might say that those folds which had their roots farther

to the south were the most drawn-out towards the north: or again we might say that the synclinal or deep-seated part of the fold has lagged behind the anticlinal or what was originally the highest part of the fold, in the advance of the latter to the north. The anticline has advanced relatively to the syncline. To this law one exception only is observed in the Swiss Alps; the sheet of the Brèche (_Byecciendecke_) falls short, in its northerly extension, of the underlying fold, which extends to form the Préalpes.

Contemplating such a generalised section as Professor Schmidt's, or, indeed, more particular sections, such as those in the Mont

Blanc Massif by Marcel Bertrand,[1] of the Dent de Morcles, Diablerets, Wildhorn, and Massif de la Brèche by Lugeon,[2] or finally Termier's section of the Pelvoux Massif,[3] one is reminded of the breaking of waves on a sloping beach. The wave, retarded at its base, is carried forward above by its momentum, and finally spreads far up on the strand; and if it could there remain, the succeeding wave must necessarily find itself superimposed upon the first. But no effects of inertia, no kinetic effects, may be called to our aid in explaining the formation of mountains. Some geologists have accordingly supposed that in order to account for

[1] Marcel Bertrand, _Cong. Géol. Internat._, 1900, Guide Géol., xiii. a, p. 41.

[2] Lugeon, _loc. cit._, p. 773.

[3] De Lapparent, _Traite de Géol._, p. 1,773.

the recumbent folds and the peculiar phenomena of increasing overlap, or _déferlement_, an obstacle, fixed and deep-seated, must have arrested the roots or synclines of the folds, and held them against translational motion, while a movement of the upper crust drew out and carried forward the anticlines. Others have contented themselves by recording the facts without advancing any explanatory hypothesis beyond that embodied in the incontestable statement that such phenomena must be referred to the effects of tangential forces acting in the Earth's crust.

It would appear that the explanation of the phenomena of recumbent folds and their _déferlement_ is to be obtained directly from the temperature conditions prevailing throughout the stressed pile of rocks; and here the subject of mountain tectonics touches that with which we were elsewhere specially concerned—the geological influence of accumulated radioactive energy.

As already shown[1], a rise of temperature due to this source of several hundred degrees might be added to such temperatures as would arise from the mere blanketing of the Earth, and the consequent upward movement of the geotherms. The time element is here the most important consideration. The whole sequence of events from the first orogenic movements to the final upheaval in Pliocene times must probably have occupied not less than ten million years.

[1] _Mountain Genesis_, p. 129, et seq.

Unfortunately the full investigation of the distribution of temperature after any given time is beset with difficulties; the conditions being extremely complex. If the radioactive heating was strictly adiabatic—that is, if all the heat was conserved and none entered from without—the time required for the attainment of the equilibrium radioactive temperature would be just about six million years. The conditions are not, indeed, adiabatic; but, on the other hand, the rocks upraised by lateral pressure were by no means at 0° C. to start with. They must be assumed to have possessed such temperatures as the prior radiothermal effects, and the conducted heat from the Earth's interior, may have established.

It would from this appear probable that if a duration of ten million years was involved, the equilibrium radioactive temperatures must nearly have been attained. The effects of heat conducted from the underlying earthcrust have to be added, leading to a further rise in temperature of not less than 500° or 600° . In such considerations the observed indications of high temperatures in materials now laid bare by denudation, probably find their explanation (P1. XIX).

The first fact that we infer from the former existence of such a temperature distribution is the improbability, indeed the impossibility, that anything resembling a rigid obstacle, or deep-seated "horst," can have existed beneath the present surface-level, and opposed the northerly movement of the deep-lying synclines. For

such a horst can only have been constituted of some siliceous rock-material such as we find everywhere rising through the worn-down sediments of the Alps; and the idea that this could retain rigidity under the prevailing temperature conditions, must be dismissed. There is no need to labour this question; the horst cannot have existed. To what, then, is the retardation of the lower parts of the folds, their overthrow, above, to the north, and their _déferlement_, to be ascribed?

A little consideration shows that the very conditions of high temperature and viscosity, which render untenable the hypothesis of a rigid obstacle, suffice to afford a full explanation of the retardation of the roots of the folds. For directed translatory movements cannot be transmitted through a fluid, pressure in which is necessarily hydrostatic, and must be exerted equally in every direction. And this applies, not only to a fluid, but to a body which will yield viscously to an impressed force. There will be a gradation, according as viscosity gives place to rigidity, between the states in which the applied force resolves itself into a purely hydrostatic pressure, and in which it is transmitted through the material as a directed thrust. The nature of the force, in the most general case, of course, has to be considered; whether it is suddenly applied and of brief duration, or steady and long-continued. The latter conditions alone apply to the present case.

It follows from this that, although a tangential force

or pressure be engendered by a crustal movement occurring to the south, and the resultant effects be transmitted northwards, these stresses can only mechanically affect the rigid parts of the crust into which they are carried. That is to say, they may result in folding and crushing, or horizontally transporting, the upper layers of the Earth's crust; but in the deeper-lying viscous materials they must be resolved into hydrostatic pressure which may act to upheave the overlying covering, but must refuse to transmit the horizontal translatory movements affecting the rigid materials above.

Between the regions in which these two opposing conditions prevail there will be no hard and fast line; but with the downward increase of fluidity there will be a gradual failure of the mechanical conditions and an increase of the hydrostatic. Thus while the uppermost layers of the crust may be transported to the full amount of the crustal displacement acting from the south (speaking still of the Alps) deeper down there will be a lesser horizontal movement, and still deeper there is no influence to urge the viscous rock-materials in a northerly direction. The consequences of these conditions must be the recumbence of the folds formed under the crust-stress, and their _déferlement_ towards the north. To see this, we must follow the several stages of development.

The earliest movements, we may suppose, result in flexures of the Jura-Mountain type—that is, in a

succession of undulations more or less symmetrical. As the orogenic force continues and develops, these undulations give place to folds, the limbs of which are approximately vertical, and the synclinal parts of which become ever more and more depressed into the deeper, and necessarily hotter, underlying materials; the anticlines being probably correspondingly elevated. These events are slowly developed, and the temperature beneath is steadily rising in consequence of the conducted interior heat, and the steady accumulation of radioactive energy in the sedimentary rocks and in the buried radioactive layer of the Earth. The work expended on the crushed and sheared rock also contributes to the developing temperature. Thus the geotherms must move upwards, and the viscous conditions extend from below; continually diminishing the downward range of the translatory movements progressing in the higher parts. While above the folded sediments are being carried northward, beneath they are becoming anchored in the growing viscosity of the medium. The anticlines will bend over, and the most southerly of the folds will gradually become pushed or bent over those lying to the north. Finally, the whole upper part of the sheaf will become horizontally recumbent; and as the uppermost folds will be those experiencing the greatest effects of the continued displacement, the _déferlement_ or overlap must necessarily arise.

We may follow these stages of mountain evolution

in a diagram (Fig. 9) in which we eliminate intermediate conditions, and regard the early and final stages of development only. In the upper sketch we suppose the lateral compression much developed and the upward movement of the geotherms in progress. The dotted line may be assumed to be a geotherm having a temperature of viscosity. If the conditions here shown persist

{Fig. 9}

indefinitely, there is no doubt that the only further developments possible are the continued crushing of the sediments and the bodily displacement of the whole mass to the north. The second figure is intended to show in what manner these results are evaded. The geotherm of viscosity has risen. All above it is affected mechanically by the continuing stress, and borne northwards in varying

degree depending upon the rigidity. The folds have been overthrown and drawn out; those which lay originally most to the south have become the uppermost; and, experiencing the maximum amount of displacement, overlap those lying beneath. There has also been a certain amount of upthrow owing to the hydrostatic pressure. This last-mentioned element of the phenomena is of highly indeterminate character, for we know not the limits to which the hydrostatic pressure may be transmitted, and where it may most readily find relief. While, according to some of the published sections, the uplifting force would seem to have influenced the final results of the orogenic movements, a discussion of its effects would not be profitable.

OTHER MINDS THAN OURS?

IN the year 1610 Galileo, looking through his telescope then newly perfected by his own hands, discovered that the planet Jupiter was attended by a train of tiny stars which went round and round him just as the moon goes round the Earth.

It was a revelation too great to be credited by mankind. It was opposed to the doctrine of the centrality of the Earth, for it suggested that other worlds constituted like ours might exist in the heavens.

Some said it was a mere optic illusion; others that he who looked through such a tube did it at the peril of his soul—it was but a delusion of Satan. Galileo converted a few of the unbelievers who had the courage to look through his telescope. To the others he said, he hoped they would see those moons on their way to heaven. Old as this story is it has never lost its pathos or its teaching.

The spirit which assailed Galileo's discoveries and which finally was potent to overshadow his declining years, closed in former days the mouths of those who asked the question written at the head of this lecture: "Are we to believe that there are other minds than ours?"

Today we consider the question in a very different spirit. Few would regard it as either foolish or improper. Its intense interest would be admitted by all, and but for the limitations closing our way on every side it would, doubtless, attract the most earnest investigation. Even on the mere balance of judgment between the probable and the improbable, we have little to go on. We know nothing definitely as to the conditions under which life may originate: whether these are such as to be rare almost to impossibility, or common almost to certainty. Only within narrow limits of temperature and in presence of certain of the elements, can life like ours exist, and outside these conditions life, if such there be, must be different from ours. Once originated it is so constituted as to assail the energies around it and to advance from less to greater. Do we know more than these vague facts? Yes, we have in our experience one other fact and one involving much.

We know that our world is very old; that life has been for many millions of years upon it; and that Man as a thinking being is but of yesterday. Here is then a condition to be fulfilled. To every world is physically assigned a limit to the period during which it is habitable according to our knowledge of life and its necessities. This limit passed and rationality missed, the chance for that world is gone for ever, and other minds than ours assuredly will not from it contemplate the universe. Looking at our own world we see that the tree of life has,

indeed, branched, leaved and, possibly, budded many times; it never bloomed but once.

All difficulties dissolve and speculations become needless under one condition only: that in which rationality may be inferred directly or indirectly by our observations on some sister world in space, This is just the evidence which in recent years has been claimed as derived from a study of the surface of Mars. To that planet our hope of such evidence is restricted. Our survey in all other directions is barred by insurmountable difficulties. Unless some meteoric record reached our Earth, revelationary of intelligence on a perished world, our only hope of obtaining such evidence rests on the observation of Mars' surface features. To this subject we confine our attention in what follows.

The observations made during recent years upon the surface features of Mars have, excusably enough, given rise to sensational reports. We must consider under what circumstances these observations have been made.

Mars comes into particularly favourable conditions for observation every fifteen years. It is true that every two years and two months we overtake him in his orbit and he is then in "opposition." That is, the Earth is between him and the sun: he is therefore in the opposite part of the heavens to the sun. Now Mars' orbit is very excentric, sometimes he is 139 million miles from the sun, and sometimes he as as much as 154 million miles from the sun. The Earth's orbit is, by comparison, almost

a circle. Evidently if we pass him when he is nearest to the sun we see him at his best; not only because he is then nearest to us, but because he is then also most brightly lit. In such favourable oppositions we are within 35 million miles of him; if Mars was in aphelion we would pass him at a distance of 61 million miles. Opposition occurs under the most favourable circumstances every fifteen years. There was one in 1862, another in 1877, one in 1892, and so on.

When Mars is 35 million miles off and we apply a telescope magnifying 1,000 diameters, we see him as if placed 35,000 miles off. This would be seven times nearer than we see the moon with the naked eye. As Mars has a diameter about twice as great as that of the moon, at such a distance he would look fourteen times the diameter of the moon. Granting favourable conditions of atmosphere much should be seen.

But these are just the conditions of atmosphere of which most of the European observatories cannot boast. It is to the honour of Schiaparelli, of Milan, that under comparatively unfavourable conditions and with a small instrument, he so far outstripped his contemporaries in the observation of the features of Mars that those contemporaries received much of his early discoveries with scepticism. Light and dark outlines and patches on the planet's surface had indeed been mapped by others, and even a couple of the canals sighted; but at the opposition of 1877 Schiaparelli first mapped any considerable

number of the celebrated "canals" and showed that these constituted an extraordinary and characteristic feature of the planet's geography. He called them "canali," meaning thereby "channels." It is remarkable indeed that a mistranslation appears really responsible for the initiation of the idea that these features are canals.

In 1882 Schiaparelli startled the astronomical world by declaring that he saw some of the canals double—that is appearing as two parallel lines. As these lines span the planet's surface for distances of many thousands of miles the announcement naturally gave rise to much surprise and, as I have said, to much scepticism. But he resolutely stuck to his statement. Here is his map of 1882. It is sufficiently startling.

In 1892 he drew a new map. It adds a little to the former map, but the doubling was not so well seen. It is just the strangest feature about this doubling that at times it is conspicuous, at times invisible. A line which is distinctly seen as a single line at one time, a few weeks later will appear distinctly to consist of two parallel lines; like railway tracks, but tracks perhaps 200 miles apart and up to 3,000 or even 4,000 miles in length.

Many speculations were, of course, made to account for the origin of such features. No known surface peculiarity on the Earth or moon at all resembles these features. The moon's surface as you know is cracked and

streaked. But the cracks are what we generally find cracks to be—either aimless, wandering lines, or, if radiating from a centre, then lines which contract in width as they leave the point of rupture. Where will we find cracks accurately parallel to one another sweeping round a planet's face with steady curvature for, 4,000 miles, and crossing each other as if quite unhampered by one another's presence? If the phenomenon on Mars be due to cracks they imply a uniformity in thickness and strength of crust, a homogeneity, quite beyond all anticipation. We will afterwards see that the course of the lines is itself further opposed to the theory that haphazard cracking of the crust of the planet is responsible for the lines. It was also suggested that the surface of the planet was covered with ice and that these were cracks in the ice. This theory has even greater difficulties than the last to contend with. Rivers have been suggested. A glance at our own maps at once disposes of this hypothesis. Rivers wander just as cracks do and parallel rivers like parallel cracks are unknown.

In time the many suggestions were put aside. One only remained. That the lines are actually the work of intelligence; actually are canals, artificially made, constructed for irrigation purposes on a scale of which we can hardly form any conception based on our own earthly engineering structures.

During the opposition of 1894, Percival Lowell, along with A. E. Douglass, and W. H. Pickering,

observed the planet from the summit of a mountain in Arizona, using an 18-inch refracting telescope and every resource of delicate measurement and spectroscopy. So superb a climate favoured them that for ten months the planet was kept under continual observation. Over 900 drawings were made and not only were Schiaparelli's channels confirmed, but they added 116 to his 79, on that portion of the planet visible at that opposition. They made the further important discovery that the lines do not stop short at the dark regions of the planet's surface, as hitherto believed, but go right on in many cases; the curvature of the lines being unaltered.

Lowell is an uncompromising advocate of the "canal" theory. If his arguments are correct we have at once an answer to our question, "Are there other minds than ours?"

We must consider a moment Lowell's arguments; not that it is my intention to combat them. You must form your own conclusions. I shall lay before you another and, as I venture to think, more adequate hypothesis in explanation of the channels of Schiaparelli. We learn, however, much from Lowell's book—it is full of interest.[1]

Lowell lays a deep foundation. He begins by showing that Mars has an atmosphere. This must be granted him till some counter observations are made.

[1] _Mars_, by Percival Lowell (Longmans, Green & Co.), 1896,

It is generally accepted. What that atmosphere is, is another matter. He certainly has made out a good case for the presence of water as one of its constituents,

It was long known that Mars possessed white regions at his poles, just as our Earth does. The waning of these polar snows—if indeed they are such—with the advance of the Martian summer, had often been observed. Lowell plots day by day this waning. It is evident from his observations that the snowfall must be light indeed. We see in his map the south pole turned towards us. Mars in perihelion always turns his south pole towards the sun and therefore towards the Earth. We see that between the dates June 3rd to August 3rd—or in two months—the polar snow had almost completely vanished. This denotes a very scanty covering. It must be remembered that Mars even when nearest to the sun receives but half our supply of solar heat and light.

But other evidence exists to show that Mars probably possesses but little water upon his surface. The dark places are not water-covered, although they have been named as if they were, indeed, seas and lakes. Various phenomena show this. The canals show it. It would never do to imagine canals crossing the seas. No great rivers are visible. There is a striking absence of clouds. The atmosphere of Mars seems as serene as that of Venus appears to be cloudy. Mists and clouds, however, sometime appear to veil his face and add to the difficulty of

making observations near the limb of the planet. Lowell concludes it must be a calm and serene atmosphere; probably only one-seventh of our own in density. The normal height of the barometer in Mars would then be but four and a half inches. This is a pressure far less than exists on the top of the highest terrestrial mountain. A mountain here must have an altitude of about ten miles to possess so low a pressure on its summit. Drops of water big enough to form rain can hardly collect in such a rarefied atmosphere. Moisture will fall as dew or frost upon the ground. The days will be hot owing to the unimpeded solar radiation; the nights bitterly cold owing to the free radiation into space.

We may add that in such a climate the frost will descend principally upon the high ground at night time and in the advancing day it will melt. The freer radiation brings about this phenomenon among our own mountains in clear and calm weather.

With the progressive melting of the snow upon the pole Lowell connected many phenomena upon the planet's surface of much interest. The dark spaces appear to grow darker and more greenish. The canals begin to show themselves and reveal their double nature. All this suggests that the moisture liberated by the melting of the polar snow with the advancing year, is carrying vitality and springtime over the surface of the planet. But how is the water conveyed?

Lowell believes principally by the canals. These are

constructed triangulating the surface of the planet in all directions. What we see, according to Lowell, is not the canal itself, but the broad band of vegetation which springs up on the arrival of the water. This band is perhaps thirty or forty miles wide, but perhaps much less, for Lowell reports that the better the conditions of observation the finer the lines appeared, so that they may be as narrow, possibly, as fifteen miles. It is to be remarked that a just visible dot on the surface of Mars must possess a diameter of 30 miles. But a chain of much smaller dots will be visible, just as we can see such fine objects as spiders' webs. The widening of the canals is then accounted for, according to Lowell, by the growth of a band of vegetation, similar to that which springs into existence when the floods of the Nile irrigate the plains of Egypt.

If no other explanation of the lines is forthcoming than that they are the work of intelligence, all this must be remembered. If all other theories fail us, much must be granted Lowell. We must not reason like fishes—as Lowell puts it—and deny that intelligent beings can thrive in an atmospheric pressure of four and half inches of mercury. Zurbriggen has recently got to the top of Aconcagua, a height of 24,000 feet. On the summit of such a mountain the barometer must stand at about ten inches. Why should not beings be developed by evolution with a lung capacity capable of living at two and a half times this altitude. Those steadily

curved parallel lines are, indeed, very unlike anything we have experience of. It would be rather to be expected that another civilisation than our own would present many wide differences in its development.

What then is the picture we have before us according to Lowell? It is a sufficiently dramatic one.

Mars is a world whose water supply, never probably very abundant, has through countless years been drying up, sinking into his surface. But the inhabitants are making a brave fight for it, They have constructed canals right round their world so that the water, which otherwise would run to waste over the vast deserts, is led from oasis to oasis. Here the great centres of civilisation are placed: their Londons, Viennas, New Yorks. These gigantic works are the works of despair. A great and civilised world finds death staring it in the face. They have had to triple their canals so that when the central canal has done its work the water is turned into the side canals, in order to utilise it as far as possible. Through their splendid telescopes they must view our seas and ample rivers; and must die like travellers in the desert seeing in a mirage the cool waters of a distant lake.

Perhaps that lonely signal reported to have been seen in the twilight limb of Mars was the outcome of pride in their splendid and perishing civilisation. They would leave some memory of it: they would have us witness how great was that civilisation before they perish!

I close this dramatic picture with the poor comfort

that several philanthropic people have suggested signalling to them as a mark of sympathy. It is said that a fortune was bequeathed to the French Academy for the purpose of communicating with the Martians. It has been suggested that we could flash signals to them by means of gigantic mirrors reflecting the light of our Sun. Or, again, that we might light bonfires on a sufficiently large scale. They would have to be about ten miles in diameter! A writer in the Pall Mall Gazette suggested that there need really be no difficulty in the matter. With the kind cooperation of the London Gas Companies (this was before the days of electric lighting) a signal might be sent without any additional expense if the gas companies would consent to simultaneously turn off the gas at intervals of five minutes over the whole of London, a signal which would be visible to the astronomers in Mars would result. He adds, naively: "If only tried for an hour each night some results might be obtained."

II

We have reviewed the theory of the artificial construction of the Martian lines. The amount of consideration we are disposed to give to the supposition that there are upon Mars other minds than ours will—as I have stated—necessarily depend upon whether or not we can assign a probable explanation of the lines upon purely physical grounds. If it is apparent that such

lines would be formed with great probability under certain conditions, which conditions are themselves probable, then the argument by exclusion for the existence of civilisation on Mars, at once breaks down.

{Fig. 10}

As a romance writer is sometimes under the necessity of transporting his readers to other scenes, so I must now ask you to consent to be transported some millions

of miles into the region of the heavens which lies outside Mars' orbit.

Between Mars and Jupiter is a chasm of 341 millions of miles. This gap in the sequence of planets was long known to be quite out of keeping with the orderly succession of worlds outward from the Sun. A society was formed at the close of the last century for the detection of the missing world. On the first day of the last century, Piazzi—who, by the way, was not a member of the society—discovered a tiny world in the vacant gap. Although eagerly welcomed, as better than nothing, it was a disappointing find. The new world was a mere rock. A speck of about 160 miles in diameter. It was obviously never intended that such a body should have all this space to itself. And, sure enough, shortly after, another small world was discovered. Then another was found, and another, and so on; and now more than 400 of these strange little worlds are known.

But whence came such bodies? The generally accepted belief is that these really represent a misbegotten world. When the Sun was younger he shed off the several worlds of our system as so many rings. Each ring then coalesced into a world. Neptune being the first born; Mercury the youngest born.

After Jupiter was thrown off, and the Sun had shrunk away inwards some 20o million miles, he shed off another ring. Meaning that this offspring of his should grow up like the rest, develop into a stable world with the

potentiality even, it may be, of becoming the abode of rational beings. But something went wrong. It broke up into a ring of little bodies, circulating around him.

It is probable on this hypothesis that the number we are acquainted with does not nearly represent the actual number of past and present asteroids. It would take 125,000 of the biggest of them to make up a globe as big as our world. They, so far as they are known, vary in size from 10 miles to 160 miles in diameter. It is probable then—on the assumption that this failure of a world was intended to be about the mass of our Earth—that they numbered, and possibly number, many hundreds of thousands.

Some of these little bodies are very peculiar in respect to the orbits they move in. This peculiarity is sometimes in the eccentricity of their orbits, sometimes in the manner in which their orbits are tilted to the general plane of the ecliptic, in which all the other planets move.

The eccentricity, according to Proctor, in some cases may attain such extremes as to bring the little world inside Mars' mean distance from the sun. This, as you will remember, is very much less than his greatest distance from the sun. The entire belt of asteroids—as known—lie much nearer to Mars than to Jupiter.

As regards the tilt of their orbits, some are actually as much as 34 degrees inclined to the ecliptic, so that in fact they are seen from the Earth among our polar constellations.

From all this you see that Mars occupies a rather hot comer in the solar system. Is it not possible that more than once in the remote past Mars may have encountered one of these wanderers? If he came within a certain distance of the small body his great mass would sway it from its orbit, and under certain conditions he would pick up a satellite in this manner. That his present satellites were actually so acquired is the suggestion of Newton, of Yale College.

Mars' satellites are indeed suspiciously and most abnormally small. I have not time to prove this to you by comparison with the other worlds of the solar system. In fact, they were not discovered till 1877—although they were predicted in a most curious manner, with the most uncannily accurate details, by Swift.

One of these bodies is about 36 miles in diameter. This is Phobos. Phobos is only 3.700 miles from the surface of Mars. The other is smaller and further off. He is named Deimos, and his diameter is only 10 miles. He is 12,500 miles from Mars' surface. With the exception of Phobos the next smallest satellite known in the solar system is one of Saturn's—Hyperion; almost 800 miles in diameter. The inner one goes all round Mars in 7½ hours. This is Phobos' month. Mars turns on his axis in 24 hours and 40 minutes, so that people in Mars would see the rise of Phobos twice in the course of a day and night; lie would apparently cross the sky

going against the other satellite; that is, he would move apparently from west to east.

We may at least assume as probable that other satellites have been gathered by Mars in the past from the army of asteroids.

Some of the satellites so picked up would be direct: that is, would move round the planet in the direction of his axial rotation. Others, on the chances, would be retrograde: that is, would move against his axial rotation. They would describe orbits making the same various angles with the ecliptic as do the asteroids; and we may be sure they would be of the same varying dimensions.

We go on to inquire what would be the consequence to Mars of such captures.

A satellite captured in this manner is very likely to be pulled into the Planet. This is a probable end of a satellite in any case. It will probably be the end of our satellite too. The satellite Phobos is indeed believed to be about to take this very plunge into his planet. But in the case when the satellite picked up happens to be rotating round the planet in the opposite direction to the axial rotation of the planet, it is pretty certain that its career as a satellite will be a brief one. The reasons for this I cannot now give. If, then, Mars picked up satellites he is very sure to have absorbed them sooner or later. Sooner if they happened to be retrograde satellites, later if direct satellites. His present satellites are recent additions. They are direct.

The path of an expiring satellite will be a slow spiral described round the planet. The spiral will at last, after many years, bring the satellite down upon the surface of the primary. Its final approach will be accelerated if the planet possesses an atmosphere, as Mars probably does. A satellite of the dimensions of Phobos—that is 36 miles in diameter—would hardly survive more than 30 to 60 years within seventy miles of Mars' surface. It will then be rotating round Mars in an hour and forty minutes, moving, in fact, at the rate of 2.2 miles per second. In the course of this 30 or 60 years it will, therefore, get round perhaps 200,000 times, before it finally crashes down upon the Martians. During this closing history of the satellite there is reason to believe, however, that it would by no means pursue continually the same path over the surface of the planet. There are many disturbing factors to be considered. Being so small any large surface features of Mars would probably act to perturb the orbit of the satellite.

The explanation of Mars' lines which I suggest, is that they were formed by the approach of such satellites in former times. I do not mean that they are lines cut into his surface by the actual infall of a satellite. The final end of the satellite would be too rapid for this, I think. But I hope to be able to show you that there is reason to believe that the mere passage of the satellite, say at 70 miles above the surface of the planet, will, in itself, give rise to effects on the crust of the planet capable

of accounting for just such single or parallel lines as we see.

In the first place we have to consider the stability of the satellite. Even in the case of a small satellite we cannot overlook the fact that the half of the satellite near the planet is pulled towards the planet by a gravitational force greater than that attracting the outer half, and that the centrifugal force is less on the inner than on the outer hemisphere. Hence there exists a force tending to tear the satellite asunder on the equatorial section tangential

{Fig. 11}

to the planet's surface. If in a fluid or plastic state, Phobos, for instance, could not possibly exist near the planet's surface. The forces referred to would decide its fate. It may be shown by calculation, however, that if Phobos has the strength of basalt or glass there would remain a considerable coefficient of safety in favour of the satellite's stability; even when the surfaces of planet and satellite were separated by only five miles.

We have now to consider some things which we expect will happen before the satellite takes its final plunge into the planet.

This diagram (Fig. 11) shows you the satellite travelling above the surface of the planet. The satellite is advancing towards, or away from, the spectator. The planet is supposed to show its solid crust in cross section, which may be a few miles in thickness. Below this is such a hot plastic magma as we have reason to believe underlies much of the solid crust of our own Earth. Now there is an attraction between the satellite and the crust of the planet; the same gravitational attraction which exists between every particle of matter in the universe. Let us consider how this attraction will affect the planet's crust. I have drawn little arrows to show how we may consider the attraction of the satellite pulling the crust of the planet not only upwards, but also pulling it inwards beneath the satellite. I have made these arrows longer where calculation shows the stress is greater. You see that the greatest lifting stress is just beneath the satellite, whereas the greatest stress pulling the crust in under the satellite is at a point which lies out from under the satellite, at a considerable distance. At each side of the satellite there is a point where the stress pulling on the crust is the greatest. Of the two stresses the lifting stress will tend to raise the crust a little; the pulling stress may in certain cases actually tear the crust across; as at A and B.

It is possible to calculate the amount of the stress at the point at each side of the satellite where the stress is at its greatest. We must assume the satellite to be a certain size and density; we must also assume the crust of

Mars to be of some certain density. To fix our ideas on these points I take the case of the present satellite Phobos. What amount of stress will he exert upon the crust of Mars when he approaches within, say, 40 miles of the planet's surface? We know his size approximately—he is about 36 miles in diameter. We can guess his density to be between four times that of water and eight times that of water. We may assume the density of Mars' surface to be about the same as that of our Earth's surface, that is three times as dense as water. We now find that the greatest stress tending to rend open the surface crust of Mars will be between 4,000 and 8,000 pounds to the square foot according to the density we assign to Phobos.

Will such a stress actually tear open the crust? We are not able to answer this question with any certainty. Much will depend upon the nature and condition of the crust. Thus, suppose that we are here (Fig. 12) looking down upon the satellite which is moving along slowly relatively to Mars' surface, in the direction of the arrow. The satellite has just passed over a weak and cracked part of the planet's crust. Here the stress has been sufficient to start two cracks. Now you know how easy it is to tear a piece of cloth when you go to the edge of it in order to make a beginning. Here the stress from the satellite has got to the edge of the crust. It is greatly concentrated just at the extremities of the cracks. It will, unler such circumstances probably carry on the

tear. If it does not do so this time, remember the satellite will some hours later be coming over the same place again, and then again for, at least, many hundreds of times. Then also we are not limited to the assumption that the

{Fig. 12}

satellite is as small as Phobos. Suppose we consider the case of a satellite approaching Mars which has a diameter double that of Phobos; a diameter still much less than that of the larger class of asteroids. Even at the distance

of 65 miles the stress will now amount to as much as from 15 to 30 tons per square foot. It is almost certain that such a stress repeated a comparatively few times over the same parts of the planet's surface would so rend the crust as to set up lines along which plutonic action would find a vent. That is, we might expect along these lines all the phenomena of upheaval and volcanic eruption which give rise to surface elevations.

The probable effect of a satellite of this dimension travelling slowly relatively to the surface of Mars is, then, to leave a very conspicuous memorial of his presence behind him. You see from the diagram that this memorial will consist o: two parallel lines of disturbance.

The linear character of the gravitational effects of the satellite is due entirely to the motion of the satellite relatively to the surface of the planet. If the satellite stood still above the surface the gravitational stress in the crust would, of course, be exerted radially outwards from the centre of the satellite. It would attain at the central point beneath the satellite its maximum vertical effect, and at some radial distance measured outwards from this point, which distance we can calculate, its maximum horizontal tearing effect. When the satellite moves relatively to the planet's crust, the horizontal tearing force acts differently according to whether it is directed in the line of motion or at right angles to this line.

In the direction of motion we see that the satellite

creates as it passes over the crust a wave of rarefaction or tension as at D, followed by compression just beneath the satellite and by a reversed direction of gravitational pull as the satellite passes onwards. These stresses rapidly replace one another as the satellite travels along. They are resisted by the inertia of the crust, and are taken up by its elasticity. The nature of this succession of alternate compressions and rarefactions in the crust possess some resemblance to those arising in an earthquake shock.

If we consider the effects taking place laterally to the line of motion we see that there are no such changes in the nature of the forces in the crust. At each passage of the satellite the horizontal tearing stress increases to a maximum, when it is exerted laterally, along the line passing through the horizontal projection of the satellite and at right angles to the line of motion, and again dies away. It is always a tearing stress, renewed again and again.

This effect is at its maximum along two particular parallel lines which are tangents to the circle of maximum horizontal stress and which run parallel with the path of the satellite. The distance separating these lines depend upon the elevation of the satellite above the planet's surface. Such lines mark out the theoretical axes of the "double canals" which future crustal movements will more fully develop.

It is interesting to consider what the effect of such

conditions would be if they arose at the surface of our own planet. We assume a horizontal force in the crust adequate to set up tensile stresses of the order, say, of fifteen tons to the square foot and these stresses to be repeated every few hours; our world being also subject to the dynamic effects we recognise in and beneath its crust.

It is easy to see that the areas over which the satellite exerted its gravitational stresses must become the foci —foci of linear form—of tectonic developments or crust movements. The relief of stresses, from whatever cause arising, in and beneath the crust must surely take place in these regions of disturbance and along these linear areas. Here must become concentrated the folding movements, which are under existing conditions brought into the geosynclines, along with their attendant volcanic phenomena. In the case of Mars such a concentration of tectonic events would not, owing to the absence of extensive subaerial denudation and great oceans, be complicated by the existence of such synclinal accumulations as have controlled terrestrial surface development. With the passage of time the linear features would probably develop; the energetic substratum continually asserting its influence along such lines of weakness. It is in the highest degree probable that radioactivity plays no less a part in Martian history than in terrestrial. The fact of radioactive heating allows us to assume the thin surface crust and continued sub-crustal energy throughout the entire period of the planet's history.

How far willl these effects resemble the double canals of Mars? In this figure and in the calculations I have given you I have supposed the satellite engaged in marking the planet's surface with two lines separated by about the interval separating the wider double canals of Mars—that is about 220 miles apart. What the distance between the lines will be, as already stated, will depend upon the height of the satellite above the surface when it comes upon a part of the crust in a condition to be affected by the stresses it sets up in it. If the satellite does its work at a point lower down above the surface the canal produced will be narrower. The stresses, too, will then be much greater. I must also observe that once the crust has yielded to the pulling stress, there is great probability that in future revolutions of the satellite a central fracture will result. For then all the pulling force adds itself to the lifting force and tends to crush the crust inwards on the central line beneath the satellite. It is thus quite possible that the passage of a satellite may give rise to triple lines. There is reason to believe that the canals on Mars are in some cases triple.

I have spoken all along of the satellite moving slowly over the surface of Mars. I have done so as I cannot at all pronounce so readily on what will happen when the satellite's velocity over the surface of Mars is very great. To account for all the lines mapped by Lowell some of them must have been produced by satellities moving relatively to the surface of Mars at velocities so great

as three miles a second or even rather more. The stresses set up are, in such cases, very difficult to estimate. It has not yet been done. Parallel lines of greatest stress or impulse ought to be formed as in the other case.

I now ask your attention to another kind of evidence that the lines are due in some way to the motion of satellites passing over the surface of Mars.

I may put the fresh evidence to which I refer, in this way: In Lowell's map (P1. XXII, p. 192), and in a less degree in Schiaparelli's map (ante p. 166), we are given the course of the lines as fragments of incomplete curves. Now these curves might have been anything at all. We must take them as they are, however, when we apply them as a test of the theory that the motion of a satellite round Mars can strike such lines. If it can be shown that satellites revolving round Mars might strike just such curves then we assume this as an added confirmation of the hypothesis.

We must begin by realising what sort of curves a satellite which disturbs the surface of a planet would leave behind it after its demise. You might think that the satellite revolving round and round the planet must simply describe a circle upon the spherical surface of the planet: a "great circle" as it is called; that is the greatest circle which can be described upon a sphere. This great circle can, however, only be struck, as you will see, when the planet is not turning upon its axis: a condition not likely to be realised.

This diagram (PI. XXI) shows the surface of a globe

covered with the usual imaginary lines of latitude and longitude. The orbit of a supposed satellite is shown by a line crossing the sphere at some assumed angle with the equator. Along this line the satellite always moves at uniform velocity, passing across and round the back of the sphere and again across. If the sphere is not turning on its polar axis then this satellite, which we will suppose armed with a pencil which draws a line upon the sphere, will strike a great circle right round the sphere. But the sphere is rotating. And it is to be expected that at different times in a planet's history the rate of rotation varies very much indeed. There is reason to believe that our own day was once only 2½ hours long, or thereabouts. After a preliminary rise in velocity of axial rotation, due to shrinkage attending rapid cooling, a planet as it advances in years rotates slower and slower. This phenomenon is due to tidal influences of the sun or of satellites. On the assumption that satellites fell into Mars there would in his case be a further action tending to shorten his day as time went on.

The effect of the rotation of the planet will be, of course, that as the satellite advances with its pencil it finds the surface of the sphere being displaced from under it. The line struck ceases to be the great circle but wanders off in another curve—which is in fact not a circle at all.

You will readily see how we find this curve. Suppose the sphere to be rotating at such a speed that while the satellite is advancing the distance _Oa_, the point _b_ on the

sphere will be carried into the path of the satellite. The pencil will mark this point. Similarly we find that all the points along this full curved line are points which will just find themselves under the satellite as it passes with its pencil. This curve is then the track marked out by the revolving satellite. You see it dotted round the back of the sphere to where it cuts the equator at a certain point. The course of the curve and the point where it cuts the equator, before proceeding on its way, entirely depend upon the rate at which we suppose the sphere to be rotating and the satellite to be describing the orbit. We may call the distance measured round the planet's equator separating the starting point of the curve from the point at which it again meets the equator, the "span" of the curve. The span then depends entirely upon the rate of rotation of the planet on its axis and of the satellite in its orbit round the planet.

But the nature of events might have been somewhat different. The satellite is, in the figure, supposed to be rotating round the sphere in the same direction as that in which the sphere is turning. It might have been that Mars had picked up a satellite travelling in the opposite direction to that in which he was turning. With the velocity of planet on its axis and of satellite in its orbit the same as before, a different curve would have been described. The span of the curve due to a retrograde satellite will be greater than that due to a direct satellite. The retrograde satellite will have a span more than half

way round the planet, the direct satellite will describe a curve which will be less than half way round the planet: that is a span due to a retrograde satellite will be more than 180 degrees, while the span due to a direct satellite will be less than 180 degrees upon the planet's equator.

I would draw your attention to the fact that what the span will be does not depend upon how much the orbit of the satellite is inclined to the equator. This only decides how far the curve marked out by the satellite will recede from the equator.

We find then, so far, that it is easy to distinguish between the direct and the retrograde curves. The span of one is less, of the other greater, than 180 degrees. The number of degrees which either sort of curve subtends upon the equator entirely depends upon the velocity of the satellite and the axial velocity of the planet.

But of these two velocities that of the satellite may be taken as sensibly invariable, when close enough to use his pencil. This depends upon the law of centrifugal force, which teaches us that the mass of the planet alone decides the velocity of a satellite in its orbit at any fixed distance from the planet's centre. The other velocity—that of the planet upon its axis—was, as we have seen, not in the past what it is now. If then Mars, at various times in his past history, picked up satellites, these satellites will describe curves round him having different spans which will depend upon the velocity of axial rotation of Mars at the time and upon this only.

In what way now can we apply this knowledge of the curves described by a satellite as a test of the lunar origin of the lines on Mars?

To do this we must apply to Lowell's map. We pick out preferably, of course, the most complete and definite curves. The chain of canals of which Acheron and Erebus are members mark out a fairly definite curve. We produce it by eye, preserving the curvature as far as possible, till it cuts the equator. Reading the span on the equator we find' it to be 255 degrees. In the first place we say then that this curve is due to a retrograde satellite. We also note on Lowell's map that the greatest rise of the curve is to a point about 32 degrees north of the equator. This gives the inclination of the satellite's orbit to the plane of Mars' equator.

With these data we calculate the velocity which the planet must have possessed at the time the canal was formed on the hypothesis that the curve was indeed the work of a satellite. The final question now remains If we determine the curve due to this velocity of Mars on its axis, will this curve fit that one which appears on Lowell's map, and of which we have really availed ourselves of only three points? To answer this question we plot upon a sphere, the curve of a satellite, in the manner I have described, assigning to this sphere the velocity derived from the span of 255 degrees. Having plotted the curve on the sphere it only remains to transfer it to Lowell's map. This is easily done.

This map (Pl. XXII) shows you the result of treating this, as well as other curves, in the manner just described. You see that whether the fragmentary curves are steep and receding far from the equator; or whether they are flat and lying close along the equator; whether they span less or more than 180 degrees; the curves determined on the supposition that they are the work of satellites revolving round Mars agree with the mapped curves; following them with wonderful accuracy; possessing their properties, and, indeed, in some cases, actually coinciding with them.

I may add that the inadmissible span of 180 degrees and spans very near this value, which are not well admissible, are so far as I can find, absent. The curves are not great circles.

You will require of me that I should explain the centres of radiation so conspicuous here and there on Lowell's map. The meeting of more than two lines at the oases is a phenomenon possibly of the same nature and also requiring explanation.

In the first place the curves to which I have but briefly referred actually give rise in most cases to nodal, or crossing points; sometimes on the equator, sometimes off the equator; through which the path of the satellite returns again and again.

These nodal points will not, however, afford a general explanation of the many-branched radiants.

It is probable that we should refer such an appearance

as is shown at the Sinus Titanum to the perturbations of the satellite's path due to the surface features on Mars. Observe that the principal radiants are situated upon the boundary of the dark regions or at the oases. Higher surface levels may be involved in both cases. Some marked difference in topography must characterise both these features. The latter may possibly originate in the destruction of satellites. Or again, they may arise in crustal disturbance of a volcanic nature, primarily induced or localised by the crossing of two canals. Whatever the origin of these features it is only necessary to assume that they represent elevated features of some magnitude to explain the multiplication of crossing lines. We must here recall what observers say of the multiplicity of the canals. According to Lowell, "What their number maybe lies quite beyond the possibility of count at present; for the better our own air, the more of them are visible."

Such innumerable canals are just what the present theory requires. An in-falling satellite will, in the course of the last 60 or 80 years of its career, circulate some 100,000 times over Mars' surface. Now what will determine the more conspicuous development of a particular canal? The mass of the satellite; the state of the surface crust; the proximity of the satellite; and the amount of repetition over the same ground. The after effects may be taken as proportional to the primary disturbance.

It is probable that elevated surface features will influence two of these conditions: the number of repetitions and the proximity to the surface. A tract 100 miles in diameter and elevated 5,000 or 10,000 feet would seriously perturb the orbit of such a body as Phobos. It is to be expected that not only would it be effective in swaying the orbit of the satellite in the horizontal direction but also would draw it down closer to the surface. It is even to be considered if such a mass might not become nodal to the satellite's orbit, so that this passed through or above this point at various inclinations with its primary direction. If acting to bring down the orbit then this will quicken the speed and cause the satellite further on its path to attain a somewhat higher elevation above the surface. The lines most conspicuous in the telescope are, in short, those which have been favoured by a combination of circumstances as reviewed above, among which crustal features have, in some cases, played a part.

I must briefly refer to what is one of the most interesting features of the Martian lines: the manner in which they appear to come and go like visions.

Something going on in Mars determines the phenomenon. On a particular night a certain line looks single. A few nights later signs of doubling are perceived, and later still, when the seeing is particularly good, not one but two lines are seen. Thus, as an example, we may take the case of Phison and Euphrates. Faint glimpses of the dual state were detected in the summer

and autumn, but not till November did they appear as distinctly double. Observe that by this time the Antarctic snows had melted, and there was in addition, sufficient time for the moisture so liberated to become diffused in the planet's atmosphere.

This increase in the definition and conspicuousness of certain details on Mars' surface is further brought into connection with the liberation of the polar snows and the diffusion of this water through the atmosphere, by the fact that the definition appeared progressively better from the south pole upwards as the snow disappeared. Lowell thinks this points to vegetation springing up under the influence of moisture; he considers, however, as we have seen, that the canals convey the moisture. He has to assume the construction of triple canals to explain the doubling of the lines.

If we once admit the canals to be elevated ranges—not necessarily of great height—the difficulty of accounting for increased definition with increase of moisture vanishes. We need not necessarily even suppose vegetation concerned. With respect to this last possibility we may remark that the colour observations, upon which the idea of vegetation is based, are likely to be uncertain owing to possible fatigue effects where a dark object is seen against a reddish background.

However this may be we have to consider what the effects of moisture increasing in the atmosphere of Mars will be with regard to the visibility of elevated ranges,

We assume a serene and rare atmosphere: the nights intensely cold, the days hot with the unveiled solar radiation. On the hill tops the cold of night will be still more intense and so, also, will the solar radiation by day. The result of this state of things will be that the moisture will be precipitated mainly on the mountains during the cold of night—in the form of frost—and during the day this covering of frost will melt; and, just as we see a heavy dew-fall darken the ground in summer, so the melting ice will set off the elevated land against the arid plains below. Our valleys are more moist than our mountains only because our moisture is so abundant that it drains off the mountains into the valleys. If moisture was scarce it would distil from the plains to the colder elevations of the hills. On this view the accentuation of a canal is the result of meteorological effects such as would arise in the Martian climate; effects which must be influenced by conditions of mountain elevation, atmospheric currents, etc. We, thus, follow Lowell in ascribing the accentuation of the canals to the circulation of water in Mars; but we assume a simple and natural mode of conveyance and do not postulate artificial structures of all but impossible magnitude. That vegetation may take part in the darkening of the elevated tracts is not improbable. Indeed we would expect that in the Martian climate these tracts would be the only fertile parts of the surface.

Clouds also there certainly are. More recent observations

appear to have set this beyond doubt. Their presence obviously brings in other possible explanations of the coming and going of elevated surface features.

Finally, we may ask what about the reliability of the maps? About this it is to be said that the most recent map—that by Lowell—has been confirmed by numerous drawings by different observers, and that it is,itself the result of over 900 drawings. It has become a standard chart of Mars, and while it would be rash to contend for absence of errors it appears certain that the trend of the principal canals may be relied on, as, also, the general features of the planet's surface.

The question of the possibility of illusion has frequently been raised. What I have said above to a great extent answers such objections. The close agreement between the drawings of different observers ought really to set the matter at rest. Recently, however, photography has left no further room for scepticism. First photographed in 1905, the planet has since been photographed many thousands of times from various observatories. A majority of the canals have been so mapped. The doubling of the canals is stated to have been also so recorded.[1]

The hypothesis which I have ventured to put before you involves no organic intervention to account for the

[1] E. C. Slipher's paper in _Popular Astronomy_ for March, 1914, gives a good account of the recent work.

details on Mars' surface. They are physical surface features. Mars presents his history written upon his face in the scars of former encounters—like the shield of Sir Launcelot. Some of the most interesting inferences of mathematical and physical astronomy find a confirmation in his history. The slowing down in the rate of axial rotation of the primary; the final inevitable destruction of the satellite; the existence in the past of a far larger number of asteroids than we at present are acquainted with; all these great facts are involved in the theory now advanced. If justifiably, then is Mars' face a veritable Principia.

To fully answer the question which heads these lectures, we should go out into the populous solitudes (if the term be permitted) which lie beyond our system. It is well that there is now no time left to do so; for, in fact, there we can only dream dreams wherein the limits of the possible and the impossible become lost.

The marvel of the infinite number of stars is not so marvellous as the rationality that fain would comprehend them. In seeking other minds than ours we seek for what is almost infinitely complex and coordinated in a material universe relatively simple and heterogeneous. In our mental attitude towards the great question, this fact must be regarded as fundamental.

I can only fitly close a discourse which has throughout weighed the question of the living thought against the unthinking laws of matter, by a paraphrase of the words

of a great poet when he, in higher and, perhaps, more philosophic language, also sought to place the one in comparison with the other.[1]

Richter thought that he was—with his human heart unstrengthened—taken by an angel among the universe of stars. Then, as they journeyed, our solar system was sunken like a faint star in the abyss, and they travelled yet further, on the wings of thought, through mightier systems: through all the countless numbers of our galaxy. But at length these also were left behind, and faded like a mist into the past. But this was not all. The dawn of other galaxies appeared in the void. Stars more countless still with insufferable light emerged. And these also were passed. And so they went through galaxies without number till at length they stood in the great Cathedral of the Universe. Endless were the starry aisles; endless the starry columns; infinite the arches and the architraves of stars. And the poet saw the mighty galaxies as steps descending to infinity, and as steps going up to infinity.

Then his human heart fainted and he longed for some narrow cell; longed to lie down in the grave that he might hide from infinity. And he said to the angel:

"Angel, I can go with thee no farther. Is there, then, no end to the universe of stars?"

[1] De Quincy in his _System of the Heavens_ gives a fine paraphrase of "Richter's Dream."

Then the angel flung up his glorious hands to the heaven of heavens, saying "End is there none to the universe of God? Lo! also there is no beginning."

THE LATENT IMAGE [1]

My inclination has led me, in spite of a lively dread of incurring a charge of presumption, to address you principally on that profound and most subtle question, the nature and mode of formation of the photographic image. I am impelled to do so, not only because the subject is full of fascination and hopefulness, but because the wide topics of photographic methods or photographic applications would be quite unfittingly handled by the president you have chosen.

I would first direct your attention to Sir James Dewar's remarkable result that the photographic plate retains considerable power of forming the latent image at temperatures approaching the absolute zero—a result which, as I submit, compels us to regard the fundamental effects progressing in the film under the stimulus of light undulations as other than those of a purely chemical nature. But few, if any, instances of chemical combination or decomposition are known at so low a temperature. Purely chemical actions cease, indeed, at far higher temperatures, fluorine being among the few bodies which still show

[1] Presidential address to the Photographic Convention of the United Kingdom, July, 1905. _Nature_, Vol. 72, p. 308.

chemical activity at the comparatively elevated temperature of -180° C. In short, this result of Sir James Dewar's suggests that we must seek for the foundations of photographic action in some physical or intra-atomic effect which, as in the case of radioactivity or fluorescence, is not restricted to intervals of temperature over which active molecular vis viva prevails. It compels us to regard with doubt the role of oxidation or other chemical action as essential, but rather points to the view that such effects must be secondary or subsidiary. We feel, in a word, that we must turn for guidance to some purely photo-physical effect.

Here, in the first place, we naturally recall the views of Bose. This physicist would refer the formation of the image to a strain of the bromide of silver molecule under the electric force in the light wave, converting it into what might be regarded as an allotropic modification of the normal bromide which subsequently responds specially to the attack of the developer. The function of the sensitiser, according to this view, is to retard the recovery from strain. Bose obtained many suggestive parallels between the strain phenomena he was able to observe in silver and other substances under electromagnetic radiation and the behaviour of the photographic plate when subjected to long-continued exposure to light.

This theory, whatever it may have to recommend it, can hardly be regarded as offering a fundamental explanation. In the first place, we are left in the dark as to what

the strain may be. It may mean many and various things. We know nothing as to the inner mechanism of its effects upon subsequent chemical actions—or at least we cannot correlate it with what is known of the physics of chemical activity. Finally, as will be seen later, it is hardly adequate to account for the varying degrees of stability which may apparently characterise the latent image. Still, there is much in Bose's work deserving of careful consideration. He has by no means exhausted the line of investigation he has originated.

Another theory has doubtless been in the minds of many. I have said we must seek guidance in some photo-physical phenomenon. There is one such which preeminently connects light and chemical phenomena through the intermediary of the effects of the former upon a component part of the atom. I refer to the phenomena of photo-electricity.

It was ascertained by Hertz and his immediate successors that light has a remarkable power of discharging negative electrification from the surface of bodies—especially from certain substances. For long no explanation of the cause of this appeared. But the electron—the ubiquitous electron—is now known with considerable certainty to be responsible. The effect of the electric force in the light wave is to direct or assist the electrons contained in the substance to escape from the surface of the body. Each electron carries away a very small charge of negative electrification. If, then, a body is

originally charged negatively, it will be gradually discharged by this convective process. If it is not charged to start with, the electrons will still be liberated at the surface of the body, and this will acquire a positive charge. If the body is positively charged at first, we cannot discharge it by illumination.

It would be superfluous for me to speak here of the nature of electrons or of the various modes in which their presence may be detected. Suffice it to say, in further connection with the Hertz effect, that when projected among gaseous molecules the electron soon attaches itself to one of these. In other words, it ionises a molecule of the gas or confers its electric charge upon it. The gaseous molecule may even be itself disrupted by impact of the electron, if this is moving fast enough, and left bereft of an electron.

We must note that such ionisation may be regarded as conferring potential chemical properties upon the molecules of the gas and upon the substance whence the electrons are derived. Similar ionisation under electric forces enters, as we now believe, into all the chemical effects progressing in the galvanic cell, and, indeed, generally in ionised solutes.

An experiment will best illustrate the principles I wish to remind you of. A clean aluminium plate, carefully insulated by a sulphur support, is faced by a sheet of copper-wire-gauze placed a couple of centimetres away from it. The gauze is maintained at a high positive

potential by this dry pile. A sensitive gold-leaf electroscope is attached to the aluminium plate, and its image thrown upon the screen. I now turn the light from this arc lamp upon the wire gauze, through which it in part passes and shines upon the aluminium plate. The electroscope at once charges up rapidly. There is a liberation of negative electrons at the surface of the aluminium; these, under the attraction of the positive body, are rapidly removed as ions, and the electroscope charges up positively.

Again, if I simply electrify negatively this aluminium plate so that the leaves of the attached electroscope diverge widely, and now expose it to the rays from the arc lamp, the charge, as you see, is very rapidly dissipated. With positive electrification of the aluminium there is no effect attendant on the illumination.

Thus from the work of Hertz and his successors we know that light, and more particularly what we call actinic light, is an effective means of setting free electrons from certain substances. In short, our photographic agent, light, has the power of expelling from certain substances the electron which is so potent a factor in most, if not in all, chemical effects. I have not time here to refer to the work of Elster and Geitel whereby they have shown that this action is to be traced to the electric force in the light wave, but must turn to the probable bearing of this phenomenon on the familiar facts of photography. I assume that the experiment I have shown you is the most

fundamental photographic experiment which it is now in our power to make.

We must first ask from what substances can light liberate electrons. There are many—metals as well as non-metals and liquids. It is a very general phenomenon and must operate widely throughout nature. But what chiefly concerns the present consideration is the fact that the haloid salts of silver are vigorously photo-electric, and, it is suggestive, possess, according to Schmidt, an activity in the descending order bromide, chloride, iodide. This is, in other words, their order of activity as ionisers (under the proper conditions) when exposed to ultra-violet light. Photographers will recognise that this is also the order of their photographic sensitiveness.

Another class of bodies also concerns our subject: the special sensitisers used by the photographer to modify the spectral distribution of sensibility of the haloid salts, _e.g._ eosine, fuchsine, cyanine. These again are electron-producers under light stimulus. Now it has been shown by Stoletow, Hallwachs, and Elster and Geitel that there is an intimate connection between photo-electric activity and the absorption of light by the substance, and, indeed, that the particular wave-lengths absorbed by the substance are those which are effective in liberating the electrons. Thus we have strong reason for believing that the vigorous photo-electric activity displayed by the special sensitisers must be dependent upon their colour absorption. You will recognise that this is just

the connection between their photographic effects and their behaviour towards light.

There is yet another suggestive parallel. I referred to the observation of Sir James Dewar as to the continued sensitiveness of the photographic film at the lowest attained extreme of temperature, and drew the inference that the fundamental photographic action must be of intra-atomic nature, and not dependent upon the vis viva of the molecule or atom. In then seeking the origin of photographic action in photo-electric phenomena we naturally ask, Are these latter phenomena also traceable at low temperatures? If they are, we are entitled to look upon this fact as a qualifying characteristic or as another link in the chain of evidence connecting photographic with photo-electric activity.

I have quite recently, with the aid of liquid air supplied to me from the laboratory of the Royal Dublin Society, tested the photo-sensibility of aluminium and also of silver bromide down to temperatures approaching that of the liquid air. The mode of observation is essentially that of Schmidt—what he terms his static method. The substance undergoing observation is, however, contained at the bottom of a thin copper tube, 5 cm. in diameter, which is immersed to a depth of about 10 cm in liquid air. The tube is closed above by a paraffin stopper which carries a thin quartz window as well as the sulphur tubes through which the connections pass. The air within is very carefully dried by phosphorus

pentoxide before the experiment. The arc light is used as source of illumination. It is found that a vigorous photo-electric effect continues in the case of the clean aluminium. In the case of the silver bromide a distinct photo-electric effect is still observed. I have not had leisure to make, as yet, any trustworthy estimate of the percentage effect at this temperature in the case of either substance. Nor have I determined the temperature accurately. The latter may be taken as roughly about -150° C,

Sir James Dewar's actual measilrements afforded twenty per cent. of the normal photographic effect at -180° C. and ten per cent. at the temperature of -252.5° C.

With this much to go upon, and the important additional fact that the electronic discharge—as from the X-ray tube or from radium—generates the latent image, I think we are fully entitled to suggest, as a legitimate lead to experiment, the hypothesis that the beginnings of photographic action involve an electronic discharge from the light-sensitive molecule; in other words that the latent image is built up of ionised atoms or molecules the result of the photo-electric effect on the illuminated silver haloid, and it is upon these ionised atoms that the chemical effects of the developer are subsequently directed. It may be that the liberated electrons ionise molecules not directly affected, or it may be that in their liberation they disrupt complex molecules built up in the ripening of the

emulsion. With the amount we have to go upon we cannot venture to particularise. It will be said that such an action must be in part of the nature of a chemical effect. This must be admitted, and, in so far as the rearrangement of molecular fabrics is involved, the result will doubtless be controlled by temperature conditions. The facts observed by Sir James Dewar support this. But there is involved a fundamental process—the liberation of the electron by the electric force in the light wave, which is a physical effect, and which, upon the hypothesis of its reality as a factor in forming the latent image, appears to explain completely the outstanding photographic sensitiveness of the film at temperatures far below those at which chemical actions in general cease.

Again, we may assume that the electron—producing power of the special sensitiser or dye for the particular ray it absorbs is responsible, or responsible in part, for the special sensitiveness it confers upon the film. Sir Wm. Abney has shown that these sensitisers are active even if laid on as a varnish on the sensitive surface and removed before development. It must be remembered, however, that at temperatures of about -50° these sensitisers lose much of their influence on the film; as I have pointed out in a paper read before the Photographic Convention of 1894.

It. appears to me that on these views the curious phenomenon of recurrent reversals does not present a problem hopeless of explanation. The process of photo-

ionisation constituting the latent image, where the ion is probably not immediately neutralised by chemical combination, presents features akin to the charging of a capacity—say a Leyden jar. There may be a rising potential between the groups of ions until ultimately a point is attained when there is a spontaneous neutralisation. I may observe that the phenomena of reversal appear to indicate that the change in the silver bromide molecule, whatever be its nature, is one of gradually increasing intensity, and finally attains a maximum when a return to the original condition occurs. The maximum is the point of most intense developable image. It is probable that the sensitiser—in this case the gelatin in which the bromide of silver is immersed—plays a part in the conditions of stability which are involved.

Of great interest in all our considerations and theories is the recent work of Wood on photographic reversal. The result of this work is—as I take it—to show that the stability of the latent image may be very various according to the mode of its formation. Thus it appears that the sort of latent effect which is produced by pressure or friction is the least stable of any. This may be reversed or wiped out by the application of any other known form of photographic stimulus. Thus an exposure to X-rays will obliterate it, or a very brief exposure to light. The latent image arising from X-rays is next in order of increasing stability. Light action will remove this. Third in order is a very brief light-shock or sudden flash. This

cannot be reversed by any of the foregoing modes of stimulation, but a long-continued undulatory stimulus, as from lamp-light, will reverse it. Last and most stable of all is the gradually built-up configuration due to long-continued light exposure. This can only be reversed by overdoing it according to the known facts of recurrent reversal. Wood takes occasion to remark that these phenomena are in bad agreement with the strain theory of Bose. We have, in fact, but the one resource—the allotropic modification of the haloid—whereby to explain all these orders of stability. It appears to me that the elasticity of the electronic theory is greater. The state of the ionised system may be very various according as it arises from continued rhythmic effects or from unorganised shocks. The ionisation due to X-rays or to friction will probably be quite unorganised, that due to light more or less stable according to the gradual and gentle nature of the forces at work. I think we are entitled to conclude that on the whole there is nothing in Wood's beautiful experiments opposed to the photo-electric origin of photographic effects, but that they rather fall in with what might be anticipated according to that theory.

When we look for further support to the views I have laid before you we are confronted with many difficulties. I have not as yet detected any electronic discharge from the film under light stimulus. This may be due to my defective experiments, or to a fact noted by Elster and Geitel concerning the photo-electric properties of gelatin.

They obtained a vigorous effect from Balmain's luminous paint, but when this was mixed in gelatin there was no external effect. Schmidt's results as to the continuance of photo-electric activity when bodies in general are dissolved in each other lead us to believe that an actual conservative property of the medium and not an effect of this on the luminous paint is here involved.

This conservative effect of the gelatin may be concerned with its efficacy as a sensitiser.

In the views I have laid before you I have endeavoured to show that the recent addition to our knowledge of the electron as an entity taking part in many physical and chemical effects should be kept in sight in seeking an explanation of the mode of origin of the latent image.[1]

[1] For a more detailed account of the subject, and some ingenious extensions of the views expressed above, see _Photo-Electricity_, by H. Stanley Allen: Longmans, Green & Ca., 1913.

PLEOCHROIC HALOES [1]

IT is now well established that a helium atom is expelled from certain of the radioactive elements at the moment of transformation. The helium atom or alpha ray leaves the transforming atom with a velocity which varies in the different radioactive elements, but which is always very great, attaining as much as 2 x 109 cms. per second; a velocity which, if unchecked, would carry the atom round the earth in less than two seconds. The alpha ray carries a positive charge of double the ionic amount.

When an alpha ray is discharged from the transforming element into a gaseous medium its velocity is rapidly checked and its energy absorbed. A certain amount of energy is thus transferred from the transforming atom to the gas. We recognise this energy in the gas by the altered properties of the latter; chiefly by the fact that it becomes a conductor of electricity. The mechanism by which this change is effected is in part known. The atoms of the gas, which appear to be freely penetrated by the alpha ray, are so far dismembered as to yield charged electrons or ions; the atoms remaining charged with an equal and opposite charge. Such a medium of

[1] Being the Huxley Lecture, delivered at the University of Birmingham on October 30th, 1912. Bedrock, Jan., 1913.

free electric charges becomes a conductor of electricity by convection when an electromotive force is applied. The gas also acquires other properties in virtue of its ionisation. Under certain conditions it may acquire chemical activity and new combinations may be formed or existing ones broken up. When its initial velocity is expended the helium atom gives up its properties as an alpha ray and thenceforth remains possessed of the ordinary varying velocity of thermal agitation. Bragg and Kleeman and others have investigated the career of the alpha ray when its path or range lies in a gas at ordinary or obtainable conditions of pressure and temperature. We will review some of the facts ascertained.

The range or distance traversed in a gas at ordinary pressures is a few centimetres. The following table, compiled by Geiger, gives the range in air at the temperature of 15° C.:

	cms.		cms.		cms.
Uranium 1 -	2.50	Thorium -	2.72	Radioactinium	4.60
Uranium 2 -	2.90	Radiothorium	3.87	Actinium X -	4.40
Ionium -	3.00	Thorium X -	4.30	Act Emanation	5.70
Radium -	3.30	Th Emanation	5.00	Actinium A -	6.50
Ra Emanation	4.16	Thorium A -	5.70	Actinium C -	5.40
Radium A -	4.75	Thorium C1 -	4.80		
Radium C -	6.94	Thorium C2 -	8.60		
Radium F -	3.77				

It will be seen that the ray of greatest range is that proceeding from thorium C2, which reaches a distance of 8.6 cms. In the uranium family the fastest ray is

that of radium C. It attains 6.94 cms. There is thus an appreciable difference between the ultimate distances traversed by the most energetic rays of the two families. The shortest ranges are those of uranium 1 and 2.

The ionisation effected by these rays is by no means uniform along the path of the ray. By examining the conductivity of the gas at different points along the path of the ray, the ionisation at these points may be determined. At the limits of the range the ionisation

{Fig. 13}

ceases. In this manner the range is, in fact, determined. The dotted curve (Fig. 13) depicts the recent investigation of the ionisation effected by a sheaf of parallel rays of radium C in air, as determined by Geiger. The range is laid out horizontally in centimetres. The numbers of ions are laid out vertically. The remarkable nature of the results will be at once apparent. We should have expected that the ray at the beginning of its path, when its velocity and kinetic energy were greatest, would have been more effective than towards the end of its range

when its energy had almost run out. But the curve shows that it is just the other way. The lagging ray, about to resign its ionising properties, becomes a much more efficient ioniser than it was at first. The maximum efficiency is, however, in the case of a bundle of parallel rays, not quite at the end of the range, but about half a centimetre from it. The increase to the maximum is rapid, the fall from the maximum to nothing is much more rapid.

It can be shown that the ionisation effected anywhere along the path of the ray is inversely proportional to the velocity of the ray at that point. But this evidently does not apply to the last 5 or 10 mms. of the range where the rate of ionisation and of the speed of the ray change most rapidly. To what are the changing properties of the rays near the end of their path to be ascribed? It is only recently that this matter has been elucidated.

When the alpha ray has sufficiently slowed down, its power of passing right through atoms, without appreciably experiencing any effects from them, diminishes. The opposing atoms begin to exert an influence on the path of the ray, deflecting it a little. The heavier atoms will deflect it most. This effect has been very successfully investigated by Geiger. It is known as "scattering." The angle of scattering increases rapidly with the decrease of velocity. Now the effect of the scattering will be to cause some of the rays to complete their ranges

or, more accurately, to leave their direct line of advance a little sooner than others. In the beautiful experiments of C. T. R. Wilson we are enabled to obtain ocular demonstration of the scattering. The photograph (Fig. 14.), which I owe to the kindness of Mr. Wilson, shows the deflection of the ray towards the end of its path. In

{Fig. 14}

this case the path of the ray has been rendered visible by the condensation of water particles under the influence of the ionisation; the atmosphere in which the ray travels being in a state of supersaturation with water vapour at the instant of the passage of the ray. It is evident that if we were observing the ionisation along a sheaf of parallel rays, all starting with equal velocity,

the effect of the bending of some of the rays near the end of their range must be to cause a decrease in the aggregate ionisation near the very end of the ultimate range. For, in fact, some of the rays complete their work of ionising at points in the gas before the end is reached. This is the cause, or at least an important contributory cause, of the decline in the ionisation near the end of the range, when the effects of a bundle of rays are being observed. The explanation does not suggest that the ionising power of any one ray is actually diminished before it finally ceases to be an alpha ray.

The full line in Fig. 13 gives the ionisation curve which it may be expected would be struck out by a single alpha ray. In it the ionisation goes on increasing till it abruptly ceases altogether, with the entire loss of the initial kinetic energy of the particle.

A highly remarkable fact was found out by Bragg. The effect of the atom traversed by the ray in checking the velocity of the ray is independent of the physical and chemical condition of the atom. He measured the "stopping power" of a medium by the distance the ray can penetrate into it compared with the distance to which it can penetrate in air. The less the ratio the greater is the stopping power. The stopping power of a substance is proportional to the square root of its atomic weight. The stopping power of an atom is not altered if it is in chemical union with another atom. The atomic weight is the one quality of importance. The physical

state, whether the element is in the solid, liquid or gaseous state, is unimportant. And when we deal with molecules the stopping power is simply proportional to the sum of the square roots of the atomic weights of the atoms entering into the molecule. This is the "additive law," and it obviously enables us to calculate what the range in any substance of known chemical composition and density will be, compared with the range in air.

This is of special importance in connection with phenomena we have presently to consider. It means that, knowing the chemical composition and density of any medium whatsoever, solid, liquid or gaseous, we can calculate accurately the distance to which any particular alpha ray will penetrate. Nor have the temperature and pressure to which the medium is subjected any influence save in so far as they may affect the proximity of one atom to another. The retardation of the alpha ray in the atom is not affected.

This valuable additive law, however, cannot be applied in strictness to the amount of ionisation attending the ray. The form of the molecule, or more generally its volume, may have an influence upon this. Bragg draws the conclusion, from this fact as well as from the notable increase of ionisation with loss of speed, that the ionisation is dependent upon the time the ray spends in the molecule. The energy of the ray is, indeed, found to be less efficient in producing ionisation in the smaller atomm.

Before leaving our review of the general laws governing the passage of alpha rays through matter, another point of interest must be referred to. We have hitherto spoken in general terms of the fact that ionisation attends the passage of the ray. We have said nothing as to the nature of the ionisation so produced. But in point of fact the ionisation due to an alpha ray is sui generis. A glance at one of Wilson's photographs (Fig. 14.) illustrates this. The white streak of water particles marks the path of the ray. The ions produced are evidently closely crowded along the track of the ray. They have been called into existence in a very minute instant of time. Now we know that ions of opposite sign if left to themselves recombine. The rate of recombination depends upon the product of the number of each sign present in unit volume. Here the numbers are very great and the volume very small. The ionic density is therefore high, and recombination very rapidly removes the ions after they are formed. We see here a peculiarity of the ionisation effected by alpha rays. It is linear in distribution and very local. Much of the ionisation in gases is again undone by recombination before diffusion leads to the separation of the ions. This "initial recombination" is greatest towards the end of the path of the ray where the ionisation is a maximum. Here it may be so effective that the form of the curve is completely lost unless a very large electromotive force is used to separate the ions when the ionisation is being investigated.

We have now reviewed recent work at sufficient length to understand something of the nature of the most important advance ever made in our knowledge of the atom. Let us glance briefly at what we have learned. The radioactive atom in sinking to a lower atomic weight casts out with enormous velocity an atom of helium. It thus loses a definite portion of its mass and of its energy. Helium which is chemically one of the most inert of the elements, is, when possessed of such great kinetic energy, able to penetrate and ionise the atoms which it meets in its path. It spends its energy in the act of ionising them, coming to rest, when it moves in air, in a few centimetres. Its initial velocity depends upon the particular radioactive element which has given rise to it. The length of its path is therefore different according to the radioactive element from which it proceeds. The retardation which it experiences in its path depends entirely upon the atomic weight of the atoms which it traverses. As it advances in its path its effectiveness in ionising the atom rapidly increases and attains a very marked maximum. In a gas the ions produced being much crowded together recombine rapidly; so rapidly that the actual ionisation may be quite concealed unless a sufficiently strong electric force is applied to separate them. Such is a brief summary of the climax of radioactive discovery:—the birth, life and death of the alpha ray. Its advent into Science has altered fundamentally our conception of

matter. It is fraught with momentous bearings upon Geological Science. How the work of the alpha ray is sometimes recorded visibly in the rocks and what we may learn from that record, I propose now to bring before you.

In certain minerals, notably the brown variety of mica known as biotite, the microscope reveals minute circular marks occurring here and there, quite irregularly. The most usual appearance is that of a circular area darker in colour than the surrounding mineral. The radii of these little disc-shaped marks when well defined are found to be remarkably uniform, in some cases four hundredths of a millimetre and in others three hundredths, about. These are the measurements in biotite. In other minerals the measurements are not quite the same as in biotite. Such minute objects are quite invisible to the naked eye. In some rocks they are very abundant, indeed they may be crowded together in such numbers as to darken the colour of the mineral containing them. They have long been a mystery to petrologists.

Close examination shows that there is always a small speck of a foreign body at the centre of the circle, and it is often possible to identify the nature of this central substance, small though it be. Most generally it is found to be the mineral zircon. Now this mineral was shown by Strutt to contain radium in quantities much exceeding those found in ordinary rock substances.

Some other mineral may occasionally form the nucleus, but we never find any which is not known to be specially likely to contain a radioactive substance. Another circumstance we notice. The smaller this central nucleus the more perfect in form is the darkened circular area surrounding it. When the circle is very perfect and the central mineral clearly defined at its centre we find by measurement that the radius of the darkened area is generally 0.033 mm. It may sometimes be 0.040 mm. These are always the measurements in biotite. In other minerals the radii are a little different.

We see in the photograph (Pl. XXIII, lower figure), much magnified, a halo contained in biotite. We are looking at a region in a rock-section, the rock being ground down to such a thickness that light freely passes through it. The biotite is in the centre of the field. Quartz and felspar surround it. The rock is a granite. The biotite is not all one crystal. Two crystals, mutually inclined, are cut across. The halo extends across both crystals, but owing to the fact that polarised light is used in taking the photograph it appears darker in one crystal than in the other. We see the zircon which composes the nucleus. The fine striated appearance of the biotite is due to the cleavage of that mineral, which is cut across in the section.

The question arises whether the darkened area surrounding the zircon may not be due to the influence of the radioactive substances contained in the zircon. The

extraordinary uniformity of the radial measurements of perfectly formed haloes (to use the name by which they have long been known) suggests that they may be the result of alpha radiation. For in that case, as we have seen, we can at once account for the definite radius as simply representing the range of the ray in biotite. The furthest-reaching ray will define the radius of the halo. In the case of the uranium family this will be radium C, and in the case of thorium it will be thorium C. Now here we possess a means of at once confirming or rejecting the view that the halo is a radioactive phenomenon and occasioned by alpha radiation; for we can calculate what the range of these rays will be in biotite, availing ourselves of Bragg's additive law, already referred to. When we make this calculation we find that radium C just penetrates 0.033 mm. and thorium C 0.040 mm. The proof is complete that we are dealing with the effects of alpha rays. Observe now that not only is the coincidence of measurement and calculation a proof of the view that alpha radiation has occasioned the halo, but it is a very complete verification of the important fact stated by Bragg, that the stopping power depends solely on the atomic weight of the atoms traversed by the ray.

We have seen that our examination of the rocks reveals only the two sorts of halo: the radium halo and the thorium halo. This is not without teaching. For why not find an actinium halo? Now Rutherford long ago suggested that this element and its derivatives were

probably an offspring of the uranium family; a side branch, as it were, in the formation of which relatively few transforming atoms took part. On Rutherford's theory then, actinium should always accompany uranium and radium, but in very subordinate amount. The absence of actinium haloes clearly supports this view. For if actinium was an independent element we would be sure to find actinium haloes. The difference in radius should be noticeable. If, on the other hand, actinium

was always associated with uranium and radium, then its effects would be submerged in those of the much more potent effects of the uranium series of elements.

It will have occurred to you already that if the radioactive origin of the halo is assured the shape of a halo is not really circular, but spherical. This is so. There is no such thing as a disc-shaped halo. The halo is a spherical volume containing the radioactive nucleus at its centre. The true radius of the halo may, therefore, only be measured on sections passing through the nucleus.

In order to understand the mode of formation of a halo we may profitably study on a diagram the events which go on within the halo-sphere. Such a diagram is seen in Fig. 15. It shows to relatively correct scale the limiting range of all the alpha-ray producing members of the uranium and thorium families. We know that each member of a family will exist in equilibrium amount within the nucleus possessing the parent element. Each alpha ray leaving the nucleus will just attain its range and then cease to affect the mica. Within the halosphere, there must be, therefore, the accumulated effects of the influences of all the rays. Each has its own sphere of influence, and the spheres are all concentric.

The radii in biotite of the several spheres are given in the following table

URANIUM FAMILY.
Radium C - 0.0330 mm.
Radium A - 0.0224 mm.
Ra Emanation - 0.0196 mm.
Radium F - 0.0177 mm.
Radium - 0.0156 mm.
Ionium - 0.0141 mm.
Uranium 1 - 0.0137 mm.
Uranium 2 - 0.0118 mm.

THORIUM FAMILY.
Thorium CE - 0.040 mm.
Thorium A - 0.026 mm.
Th Emanation - 0.023 mm.
Thorium Ci - 0.022 mm.
Thorium X - 0.020 mm.
Radiothorium - 0.119 mm.
Thorium - 0.013 mm.

In the photograph (Pl. XXIV, lower figure), we see a uranium and a thorium halo in the same crystal of mica. The mica is contained in a rock-section and is cut across the cleavage. The effects of thorium Ca are clearly shown

as a lighter border surrounding the accumulated inner darkening due to the other thorium rays. The uranium halo (to the right) similarly shows the effects of radium C, but less distinctly.

Haloes which are uniformly dark all over as described above are, in point of fact, "over-exposed"; to borrow a familiar photographic term. Haloes are found which show much beautiful internal detail. Too vigorous action obscures this detail just as detail is lost in an over-exposed photograph. We may again have "under-exposed" haloes in which the action of the several rays is incomplete or in which the action of certain of the rays has left little if any trace. Beginning at the most under-exposed haloes we find circular dark marks having the radius 0.012 or 0.013 mm. These haloes are due to uranium, although their inner darkening is doubtless aided by the passage of rays which were too few to extend the darkening beyond the vigorous effects of the two uranium rays. Then we find haloes carried out to the radii 0.016, 0.018 and 0.019 mm. The last sometimes show very beautiful outer rings having radial dimensions such as would be produced by radium A and radium C. Finally we may have haloes in which interior detail is lost so far out as the radius due to emanation or radium A, while outside this floats the ring due to radium C. Certain variations of these effects may occur, marking, apparently, different stages of exposure. Plates XXIII and XXIV (upper figure) illustrate some of these stages;

the latter photograph being greatly enlarged to show clearly the halo-sphere of radium A.

In most of the cases mentioned above the structure evidently shows the existence of concentric spherical shells of darkened biotite. This is a very interesting fact. For it proves that in the mineral the alpha ray gives rise to the same increased ionisation towards the end of its range, as Bragg determined in the case of gases. And we must conclude that the halo in every case grows in this manner. A spherical shell of darkened biotite is first produced and the inner colouration is only effected as the more feeble ionisation along the track of the ray in course of ages gives rise to sufficient alteration of the mineral. This more feeble ionisation is, near the nucleus, enhanced in its effects by the fact that there all the rays combine to increase the ionisation and, moreover, the several tracks are there crowded by the convergency to the centre. Hence the most elementary haloes seldom show definite rings due to uranium, etc., but appear as embryonic disc-like markings. The photographs illustrate many of the phases of halo development.

Rutherford succeeded in making a halo artificially by compressing into a capillary glass tube a quantity of the emanation of radium. As the emanation decayed the various derived products came into existence and all the several alpha rays penetrated the glass, darkening the walls of the capillary out to the limit of the range of radium C in glass. Plate XXV shows a magnified section of the tube. The dark central part is the capillary. The tubular halo surrounds it. This experiment has, however, been anticipated by some scores of millions of years, for here is the same effect in a biotite crystal (Pl. XXV). Along what are apparently tubular passages or cracks in the mica, a solution, rich in radioactive substances, has moved; probably during the final consolidation of the granite in which the mica occurs. A continuous and very regular halo has developed along these conduits. A string of halo-spheres may lie along such passages. We must infer that solutions or gases able to establish the radioactive nuclei moved along these conduits, and we are entitled to ask if all the haloes in this biotite are not, in this sense, of secondary origin. There is, I may add, much to support such a conclusion.

The widespread distribution of radioactive substances is most readily appreciated by examination of sections of rocks cut thin enough for microscopic investigation. It is, indeed, difficult to find, in the older rocks of granitic type, mica which does not show haloes, or traces of haloes. Often we find that every one of the inclusions in the mica—that is, every one of the earlier formed substances—contain radioactive elements, as indicated by the presence of darkened borders. As will be seen presently the quantities involved are generally vanishingly small. For example it was found by direct determination that in one gram of the halo-rich mica of Co. Carlow there was rather less than twelve billionths of a gram of radium, We are

entitled to infer that other rare elements are similarly widely distributed but remain undetectable because of their more stable properties.

It must not be thought that the under-exposed halo is a recent creation. By no means. All are old, appallingly old; and in the same rock all are, probably, of the same, or neatly the same, age. The under-exposure is simply due to a lesser quantity of the radioactive elements in the nucleus. They are under-exposed, in short, not because of lesser duration of exposure, but because of insufficient action; as when in taking a photograph the stop is not open enough for the time of the exposure.

The halo has, so far, told us that the additive law is obeyed in solid media, and that the increased ionisation attending the slowing down of the ray obtaining in gases, also obtains in solids; for, otherwise, the halo would not commence its development as a spherical shell or envelope. But here we learn that there is probably a certain difference in the course of events attending the immediate passage of the ray in the gas and in the solid. In the former, initial recombination may obscure the intense ionisation near the end of the range. We can only detect the true end-effects by artificially separating the ions by a strong electric force. If this recombination happened in the mineral we should not have the concentric spheres so well defined as we see them to be. What, then, hinders the initial recombination in the solid? The answer probably is that the newly formed

ion is instantly used up in a fresh chemical combination. Nor is it free to change its place as in the gas. There is simply a new equilibrium brought about by its sudden production. In this manner the conditions in the complex molecule of biotite, tourmaline, etc., may be quite as effective in preventing initial recombination as the most effective electric force we could apply. The final result is that we find the Bragg curve reproduced most accurately in the delicate shading of the rings making up the perfectly exposed halo.

That the shading of the rings reproduces the form of the Bragg curve, projected, as it were, upon the line of advance of the ray and reproduced in depth of shading, shows that in yet another particular the alpha ray behaves much the same in the solid as in the gas. A careful examination of the outer edge of the circles always reveals a steep but not abrupt cessation of the action of the ray. Now Geiger has investigated and proved the existence of scattering of the alpha ray by solids. We may, therefore, suppose with much probability that there is the same scattering within the mineral near the end of the range. The heavy iron atom of the biotite is, doubtless, chiefly responsible for this in biotite haloes. I may observe that this shading of the outer bounding surface of the sphere of action is found however minute the central nucleus. In the case of a nucleus of considerable size another effect comes in which tends to produce an enhanced shading. This will

result if rays proceed from different depths in the nucleus. If the nucleus were of the same density and atomic weight as the surrounding mica, there would be little effect. But its density and molecular weight are generally greater, hence the retardation is greater, and rays proceeding from deep in the nucleus experience more retardation than those which proceed from points near to the surface. The distances reached by the rays in the mica will vary accordingly, and so there will be a gradual cessation of the effects of the rays.

The result of our study of the halo may be summed up in the statement that in nearly every particular we have the phenomena, which have been measured and observed in the gas, reproduced on a minute scale in the halo. Initial recombination seems, however, to be absent or diminished in effectiveness; probably because of the new stability instantly assumed by the ionised atoms.

One of the most interesting points about the halo remains to be referred to. The halo is always uniformly darkened all round its circumference and is perfectly spherical. Sections, whether taken in the plane of cleavage of the mica or across it, show the same exactly circular form, and the same radius. Of course, if there was any appreciable increase of range along or across the cleavage the form of the halo on the section across the cleavage should be elliptical. The fact that there is no measurable ellipticity is, I think, one which on first consideration would not be expected.

For what are the conditions attending the passage of the ray in a medium such as mica? According to crystallographic conceptions we have here an orderly arrangement of molecules, the units composing the crystal being alike in mass, geometrically spaced, and polarised as regards the attractions they exert one upon another. Mica, more especially, has the cleavage phenomenon developed to a degree which transcends its development in any other known substance. We can cleave it and again cleave it till its flakes float in the air, and we may yet go on cleaving it by special means till the flakes no longer reflect visible light. And not less remarkable is the uniplanar nature of its cleavage. There is little cleavage in any plane but the one, although it is easy to show that the molecules in the plane of the flake are in orderly arrangement and are more easily parted in some directions than in others. In such a medium beyond all others we must look with surprise upon the perfect sphere struck out by the alpha rays, because it seems certain that the cleavage is due to lesser attraction, and, probably, further spacing of the molecules, in a direction perpendicular to the cleavage.

It may turn out that the spacing of the molecules will influence but little the average number per unit distance encountered by rays moving in divergent paths. If this is so, we seem left to conclude that, in spite of its unequal and polarised attractions, there is equal retardation and equal ionisation in the molecule in whatever

direction it is approached. Or, again, if the encounters indeed differ in number, then some compensating effect must exist whereby a direction of lesser linear density involves greater stopping power in the molecule encountered, and vice versa.

The nature of the change produced by the alpha rays is unknown. But the formation of the halo is not, at least in its earlier stages, attended by destruction of the crystallographic and optical properties of the medium. The optical properties are unaltered in nature but are increased in intensity. This applies till the halo has become so darkened that light is no longer transmitted under the conditions of thickness obtaining in rock sections. It is well known that there is in biotite a maximum absorption of a plane-polarised light ray, when the plane of vibration coincides with the plane of cleavage. A section across the cleavage then shows a maximum amount of absorption. A halo seen on this section simply produces this effect in a more intense degree. This is well shown in Plate XXIII (lower figure), on a portion of the halo-sphere. The descriptive name "Pleochroic Halo" has originated from this fact. We must conclude that the effect of the ionisation due to the alpha ray has not been to alter fundamentally the conditions which give rise to the optical properties of the medium. The increased absorption is probably associated with some change in the chemical state of the iron present. Haloes are, I believe, not found in minerals from which this

element is absent. One thing is quite certain. The colouration is not due to an accumulation of helium atoms, _i.e._ of spent alpha rays. The evidence for this is conclusive. If helium was responsible we should have haloes produced in all sorts of colourless minerals. Now we sometimes see zircons in felspars and in quartz, etc., but in no such case is a halo produced. And halo-spheres formed within and sufficiently close to the edge of a crystal of mica are abruptly truncated by neighbouring areas of felspar or quartz, although we know that the rays must pass freely across the boundary. Again it is easy to show that even in the oldest haloes the quantity of helium involved is so small that one might say the halo-sphere was a tolerably good vacuum as regards helium. There is, finally, no reason to suppose that the imprisoned helium would exhibit such a colouration, or, indeed, any at all.

I have already referred to the great age of the halo. Haloes are not found in the younger igneous rocks. It is probable that a halo less than a million years old has never been seen. This, primâ facie, indicates an extremely slow rate of formation. And our calculations quite support the conclusions that the growth of a halo, if this has been uniform, proceeds at a rate of almost unimaginable slowness.

Let us calculate the number of alpha rays which may have gone to form a halo in the Devonian granite of Leinster.

It is common to find haloes developed perfectly in this granite, and having a nucleus of zircon less than 5×10^{-4} cms. in diameter. The volume of zircon is 65×10^{-12} c.cs. and the mass 3×10^{-10} grm.; and if there was in this zircon 10^{-8} grm. radium per gram (a quantity about five times the greatest amount measured by Strutt), the mass of radium involved is 3×10^{-18} grm. From this and from the fact ascertained by Rutherford that the number of alpha rays expelled by a gram of radium in one second is 3.4×10^{10}, we find that three rays are shot from the nucleus in a year. If, now, geological time since the Devonian is 50 millions of years, then 150 millions of rays built up the halo. If geological time since the Devonian is 400 millions of years, then 1,200 millions of alpha rays are concerned in its genesis. The number of ions involved, of course, greatly exceeds these numbers. A single alpha ray fired from radium C will produce 2.37×10^{5} ions in air.

But haloes may be found quite clearly defined and fairly dark out to the range of the emanation ray and derived from much less quantities of radioactive materials. Thus a zircon nucleus with a diameter of but 3.4×10^{-4} cms. formed a halo strongly darkened within, and showing radium A and radium C as clear smoky rings.

Such a nucleus, on the assumption made above as to its radium content, expels one ray in a year. But, again, haloes are observed with less blackened pupils and with faint ring due to radium C, formed round nuclei

of rather less than 2×10^{-4} cms. diameter. Such nuclei would expel one ray in five years. And even lesser nuclei will generate in these old rocks haloes with their earlier characteristic features clearly developed. In the case of the most minute nuclei, if my assumption as to the uranium content is correct, an alpha ray is expelled, probably, no oftener than once in a century; and possibly at still longer intervals.

The equilibrium amount of radium contained in some nuclei may amount to only a few atoms. Even in the case of the larger nuclei and more perfectly developed haloes the quantity of radium involved is many millions of times less than the least amount we can recognise by any other means. But the delicacy of the observation is not adequately set forth in this statement. We can not only tell the nature of the radioactive family with which we are dealing; but we can recognise the presence of some of its constituent members. I may say that it is not probable the zircons are richer in radium than I have assumed. My assumption involves about 3 per cent. of uranium. I know of no analyses ascribing so great an amount of uranium to zircon. The variety cyrtolite has been found to contain half this amount, about. But even if we doubled our estimate of radium content, the remarkable nature of our conclusions is hardly lessened.

It may appear strange that the ever-interesting question of the Earth's age should find elucidation from the

study of haloes. Nevertheless the subjects are closely connected. The circumstances are as follows. Geologists have estimated the age of the Earth since denudation began, by measurements of the integral effects of denudation. These methods agree in showing an age of about rob years. On the other hand, measurements have been made of the accumulation in minerals of radioactive _débris_—the helium and lead—and results obtained which, although they do not agree very well among themselves, are concordant in assigning a very much greater age to the rocks. If the radioactive estimate is correct, then we are now living in a time when the denudative forces of the Earth are about eight or nine times as active as they have been on the average over the past. Such a state of things is absolutely unaccountable. And all the more unaccountable because from all we know we would expect a somewhat _lesser_ rate of solvent denudation as the world gets older and the land gets more and more loaded with the washed-out materials of the rocks.

Both the methods referred to of finding the age assume the principle of uniformity. The geologist contends for uniformity throughout the past physical history of the Earth. The physicist claims the like for the change-rates of the radioactive elements. Now the study of the rocks enables us to infer something as to the past history of our Globe. Nothing is, on the other hand, known respecting the origin of uranium or thorium—the parent radioactive bodies. And while not questioning the law

and regularity which undoubtedly prevail in the periods of the members of the radioactive families, it appears to me that it is allowable to ask if the change rate of uranium has been always what we now believe it to be. This comes to much the same thing as supposing that atoms possessing a faster change rate once were associated with it which were capable of yielding both helium and lead to the rocks. Such atoms might have been collateral in origin with uranium from some antecedent element. Like helium, lead may be a derivative from more than one sequence of radioactive changes. In the present state of our knowledge the possibilities are many. The rate of change is known to be connected with the range of the alpha ray expelled by the transforming element; and the conformity of the halo with our existing knowledge of the ranges is reason for assuming that, whatever the origin of the more active associate of uranium, this passed through similar elemental changes in the progress of its disintegration. There may, however, have been differences in the ranges which the halo would not reveal. It is remarkable that uranium at the present time is apparently responsible for two alpha rays of very different ranges. If these proceed from different elements, one should be faster in its change rate than the other. Some guidance may yet be forthcoming from the study of the more obscure problems of radioactivity.

Now it is not improbable that the halo may contribute directly to this discussion. We can evidently attack the biotite with a known number of alpha rays and determine how many are required to produce a certain intensity of darkening, corresponding to that of a halo with a nucleus of measurable dimensions. On certain assumptions, which are correct within defined limits, we can calculate, as I have done above, the number of rays concerned in forming the halo. In doing so we assume some value for the age of the halo. Let us take the maximum radioactive value. A halo originating in Devonian times may attain a certain central blackening from the effects of, say, rob rays. But now suppose we find that we cannot produce the same degree of blackening with this number of rays applied in the laboratory. What are we to conclude? I think there is only the one conclusion open to us; that some other source of alpha rays, or a faster rate of supply, existed in the past. And this conclusion would explain the absence of haloes from the younger rocks; which, in view of the vast range of effects possible in the development of haloes, is, otherwise, not easy to account for. It is apparent that the experiment on the biotite has a direct bearing on the validity of the radioactive method of estimating the age of the rocks. It is now being carried out by Professor Rutherford under reliable conditions.

Finally, there is one very certain and valuable fact to be learned from the halo. The halo has established the extreme rarity of radioactivity as an atomic phenomenon. One and all of the speculations as to

the slow breakdown of the commoner elements may be dismissed. The halo shows that the mica of the rocks is radioactively sensitive. The fundamental criterion of radioactive change is the expulsion of the alpha ray. The molecular system of the mica and of many other minerals is unstable in presence of these rays, just as a photographic plate is unstable in presence of light. Moreover, the mineral integrates the radioactive effects in the same way as a photographic salt integrates the effects of light. In both cases the feeblest activities become ultimately apparent to our inspection. We have seen that one ray in each year since the Devonian period will build the fully formed halo: an object unlike any other appearance in the rocks. And we have been able to allocate all the haloes so far investigated to one or the other of the known radioactive families. We are evidently justified in the belief that had other elements been radioactive we must either find characteristic haloes produced by them, or else find a complete darkening of the mica. The feeblest alpha rays emitted by the relatively enormous quantities of the prevailing elements, acting over the whole duration of geological time—and it must be remembered that the haloes we have been studying are comparatively young—must have registered their effects on the sensitive minerals. And thus we are safe in concluding that the common elements, and, indeed, many which would be called rare, are possessed of a degree of stability which has preserved them un

changed since the beginning of geological time. Each unaffected flake of mica is, thus, unassailable proof of a fact which but for the halo would, probably, have been for ever beyond our cognisance.

THE USE OF RADIUM IN MEDICINE [1]

IT has been unfortunate for the progress of the radioactive treatment of disease that its methods and claims involve much of the marvellous. Up till recently, indeed, a large part of radioactive therapeutics could only be described as bordering on the occult. It is not surprising that when, in addition to its occult and marvellous characters, claims were made on its behalf which in many cases could not be supported, many medical men came to regard it with a certain amount of suspicion.

Today, I believe, we are in a better position. I think it is possible to ascribe a rational scientific basis to its legitimate claims, and to show, in fact, that in radioactive treatment we are pursuing methods which have been already tried extensively and found to be of definite value; and that new methods differ from the old mainly in their power and availability, and little, or not at all, in kind.

Let us briefly review the basis of the science. Radium is a metallic element chemically resembling barium. It

[1] A Lecture to Postgraduate Students of Medicine in connection with the founding of the Dublin Radium Institute, delivered in the School of Physic in Ireland, Trinity College, on October 2nd, 1914

possesses, however, a remarkable property which barium does not. Its atoms are not equally stable. In a given quantity of radium a certain very small percentage of the total number of atoms present break up per second. By "breaking up" we mean their transmutation to another element. Radium, which is a solid element under ordinary conditions, gives rise by transmutation to a gaseous element—the emanation of radium. The new element is a heavy gas at ordinary temperatures and, like other gases, can be liquified by extreme cold. The extraordinary property of transmutation is entirely automatic. No influence which chemist or physicist can apply can affect the rate of transformation.

The emanation inherits the property of instability, but in its case the instability is more pronounced. A relatively large fraction of its atoms transmute per second to a solid element designated Radium A. In turn this new generation of atoms breaks up—even faster than the emanation—becoming yet another element with specific chemical properties. And so on for a whole sequence of transmutations, till finally a stable substance is formed, identical with ordinary lead in chemical and physical properties, but possessing a slightly lower atomic weight.

The genealogy of the radium series of elements shows that radium is not the starting point. It possesses ancestors which have been traced back to the element uranium.

Now what bearing has this series of transmutations

upon medical science? Radium or emanation, &c., are not in the Pharmacopoeia as are, say, arsenic or bismuth. The whole medicinal value of these elements resides in the very wonderful phenomena of their radiations. They radiate in the process of transmuting.

The changing atom may radiate a part of its own mass. The "alpha"-ray (a-ray) is such a material ray. It is an electrified helium atom cast out of the parent atom with enormous velocity—such a velocity as would carry it, if not impeded, all round the earth in two seconds. All alpha-rays are positively electrified atoms of the element helium, which thereby is shown to be an integral constituent of many elements. The alpha-ray is not of much value to medical science, for, in spite of its great velocity, it is soon stopped by encounter with other atoms. It can penetrate only a minute fraction of a millimetre into ordinary soft tissues. We shall not further consider it.

Transmuting atoms give out also material rays of another kind: the ß-rays. The ß-ray is in mass but a very small fraction of, even, a hydrogen atom. Its speed may approach that of light. As cast out by radioactive elements it starts with speeds which vary with the element, and may be from one-third to nine-tenths the velocity of light. The ß-ray is negatively electrified. It has long been known to science as the electron. It is also identical with the cathode ray of the vacuum tube.

Another and quite different kind of radiation is given out by many of the transmuting elements:—the y-ray. This is not material, it is ethereal. It is known now with certainty that the y-ray is in kind identical with light, but of very much shorter wave length than even the extreme ultraviolet light of the solar spectrum. The y-ray is flashed from the transmuting atom along with the ß-ray. It is identical in character with the x-ray but of even shorter wave length.

There is a very interesting connection between the y-ray and the ß-ray which it is important for the medical man to understand—as far as it is practicable on our present knowledge.

When y-rays or x-rays fall on matter they give rise to ß-rays. The mechanism involved is not known but it is possibly a result of the resonance of the atom, or of parts of it, to the short light waves. And it is remarkable that the y-rays which, as we have seen, are shorter and more penetrating waves than the x-rays, give rise to ß-rays possessed of greater velocity and penetration than ß-rays excited by the x-rays. Indeed the ß-rays originated by y-rays may attain a velocity nearly approaching that of light and as great as that of any ß-rays emitted by transmuting atoms. Again there is demonstrable evidence that ß-rays impinging on matter may give rise to y-rays. The most remarkable demonstration of this is seen in the x-ray tube. Here the x-rays originate where the stream of ß- or cathode-rays

are arrested on the anode. But the first relation is at present of most importance to us—_i.e._ that the y-or x-rays give rise to ß-rays.

This relation gives us additional evidence of the identity of the physical effects of y-, x-, and light-rays —using the term light rays in the usual sense of spectral rays. For it has long been known that light waves liberate electrons from atoms. It has been found that these electrons possess a certain initial velocity which is the greater the shorter the wave length of the light concerned in their liberation. The whole science of "photo-electricity" centres round this phenomenon. The action of light on the photographic plate, as well as many other physical and chemical phenomena, find an explanation in this liberation of the electron by the light wave.

Here, then, we have spectral light waves liberating electrons—_i.e._ very minute negatively-charged particles, and we find that, as we use shorter light waves, the initial velocity of these particles increases. Again, we have x-rays which are far smaller in wave length than spectral light, liberating much faster negatively electrified particles. Finally, we have y-rays—the shortest nether waves of all-liberating negative particles of the highest velocity known. Plainly the whole series of phenomena is continuous.

We can now look closer at the actions involved in the therapeutic influence of the several rays and in

this way, also, see further the correlation between what may be called photo-therapeutics and radioactive therapeutics.

The ß-ray, whether we obtain it directly from the transforming radioactive atom or whether we obtain it as a result of the effects of the y- or x-rays upon the atom, is an ionising agent of wonderful power. What is meant by this? In its physical aspect this means that the atoms through which it passes acquire free electric charges; some becoming positive, some negative. This can only be due to the loss of an electron by the affected atom. The loss of the small negative charge carried in the electron leaves the atom positively electrified or creates a positive ion. The fixing of the wandering electron to a neutral atom creates a negative ion. Before further consideration of the importance of the phenomenon of ionisation we must fix in our minds that the agent, which brings this about, is the ß-ray. There is little evidence that the y-ray can directly create ions to any large extent. But the action of liberating high-speed ß-rays results in the creation of many thousands of ions by each ß-ray liberated.

As an agent in the hands of the medical man we must regard the y-ray as a light wave of extremely penetrating character, which creates high-speed ß-rays in the tissues which it penetrates, these ß-rays being most potent ionising agents. The ß-rays directly obtained from radioactive atoms assist in the work of ionisation. ß-rays do not

penetrate far from their source. The fastest of them would not probably penetrate one centimetre in soft tissues.

We must now return to the phenomenon of ionisation. Ionisation is revealed to observation most conspicuously when it takes place in a gas. The + and - electric charges on the gas particles endow it with the properties of a conductor of electricity, the + ions moving freely in one direction and the - ions in the opposite direction under an electric potential. But there are effects brought about by ionisation of more importance to the medical man than this. The chemist has long come to recognise that in the ion he is concerned with the inner mechanism of a large number of chemical phenomena. For with the electrification of the atom attractive and repulsive forces arise. We can directly show the chemical effects of the ionising ß-rays. Water exposed to their bombardment splits up into hydrogen and oxygen. And, again, the separated atoms may be in part recombined under the influence of the radiation. Ammonia splits up into hydrogen and nitrogen. Carbon dioxide forms carbon, carbon monoxide, and oxygen; hydrochloric acid forms chlorine and hydrogen. In these cases, also, recombination can be partially effected by the rays.

We can be quite sure that within the complex structure of the living cell the ionising effects which everywhere accompany the ß-rays must exert a profound influence. The sequence of chemical events which as yet seem

beyond the ken of science and which are involved in metabolism cannot fail to be affected. Any, it is not surprising that as the result of eaperinient it is found that the radiations are agents which may be used either for the stimulation of the natural events of growth or used for the actual destruction of the cell. It is easy to see that the feeble radiation should produce the one effect, the strong the other. In a similar way by a moderate light stimulus we create the latent image in the photographic plate; by an intense light we again destroy this image. The inner mechanism in this last case can be logically stated.[1]

There is plainly a true physical basis here for the efficacy of radioactive treatment and, what is more, we find when we examine it, that it is in kind not different from that underlying treatment by spectral radiations. But in degree it is very different and here is the reason for the special importance of radioactivity as a therapeutic agent. The Finsen light is capable of influencing the soft tissues to a short depth only. The reason is that the wave length of the light used is too great to pass without rapid absorption through the tissues; and, further, the electrons it gives rise to—_i.e._ the ß-rays it liberates—are too slow-moving to be very efficient ionisers. X-rays penetrate in some cases quite freely and give rise to much faster and more powerful ß-rays

[1] See _The Latent Image_, p. 202.

than can the Finsen light. But far more penetrating than x-rays are the y-rays emitted in certain of the radioactive changes. These give rise to ß-rays having a velocity approximate to that of light.

The y-rays are, therefore, very penetrating and powerfully ionising light waves; light waves which are quite invisible to the eye and can beam right through the tissues of the body. To the mind's eye only are they visible. And a very wonderful picture they make. We see the transmuting atom flashing out this light for an inconceivably short instant as it throws off the ß-ray. And "so far this little candle throws his beams" in the complex system of the cells, so far atoms shaken by the rays send out ß-rays; these in turn are hurled against other atomic systems; fresh separations of electrons arise and new attractions and repulsions spring up and the most important chemical changes are brought about. Our mental picture can claim to be no more than diagrammatic of the reality. Still we are here dealing with recognised physical and chemical phenomena, and their description as "occult" in the derogatory sense is certainly not justifiable.

Having now briefly reviewed the nature of the rays arising in radioactive substances and the rationale of their influence, we must turn to more especially practical considerations.

The Table given opposite shows that radium itself is responsible for a- and ß-rays only. It happens that

Period in whioh ½ element is transformed.

URANIUM 1 & 2 { a 6 } x 10^9 years.

URANIUM X { a ß } 24.6 days.

IONIUM { a 8 } x 104 years.

RADIUM { a ß } 2 x 10^2 years.

EMANATION { a } 8.85 days.

RADIUM A { a 8 } minutes.

RADIUM B { ß y } 26.7 minutes.

RADIUM C { a ß y } 13.5 minutes.

RADIUM D { ß } 15 years.

RADIUM E { ß y } 4.8 days.

RADIUM (Polonium) F { a } 140 days.

Table showing the successive generations of the elements of the Uranium-radium family, the character of their radiations and their longevity.

the ß-rays emitted by radium are very "soft"—_i.e._ slow and easily absorbed. The a-ray is in no case available for more than mere surface application. Hence we see that, contrary to what is generally believed, radium itself is of little direct therapeutic value. Nor is the next body in succession—the emanation, for it gives only a-rays. In fact, to be brief, it is not till we come to Radium B that ß-rays of a relatively high penetrative quality are reached; and it is not till we come to Radium C that highly penetrative y-rays are obtained.

It is around this element, Radium C, that the chief medical importance of radioactive treatment by this family of radioactive bodies centres. Not only are ß-rays of Radium C very penetrating, but the y-rays are perhaps the most energetic rays of the, kind known. Further in the list there is no very special medical interest.

Now, how can we get a supply of this valuable element Radium C? We can obtain it from radium itself. For even if radium has been deprived of its emanation (which is easily done by heating it or bringing it into solution) in a few weeks we get back the Radium C. One thing here we must be clear about. With a given quantity of Radium only a certain definitely limited amount of Radium C, or of emanation, or any other of the derived bodies, will be associated. Why is this? The answer is because the several successive elements are themselves decaying —_i.e._ changing one into the other. The atomic per-

centage of each, which decays in a second, is a fixed quantity which we cannot alter. Now if we picture radium which has been completely deprived of its emanation, again accumulating by automatic transmutation a fresh store of this element, we have to remember:— (i) That the rate of creation of emanation by the radium is practically constant; and (2) that the absolute amount of the emanation decaying per second increases as the stock of emanation increases. Finally, when the amount of accumulated emanation has increased to such an extent that the number of emanation atoms transmuting per second becomes exactly equal to the number being generated per second, the amount of emanation present cannot increase. This is called the equilibrium amount. If fifteen members are elected steadily each year into a newly-founded society the number of members will increase for the first few years; finally, when the losses by death of the members equal about fifteen per annum the society can get no bigger. It has attained the equilibrium number of members.

This applies to every one of the successive elements. It takes twenty-one days for the equilibrium quantity of emanation to be formed in radium which has been completely de-emanated; and it takes 3.8 days for half the equilibrium amount to be formed. Again, if we start with a stock of emanation it takes just three hours for the equilibrium amount of Radium C to be formed.

We can evidently grow Radium C either from radium itself or from the emanation of radium. If we use a tube of radium we have an almost perfectly constant quantity of Radium C present, for as fast as the Radium C and intervening elements decay the Radium, which only diminishes very slowly in amount, makes up the loss. But, if we start off with a tube of emanation, we do not possess a constant supply of Radium C, because the emanation is decaying fairly rapidly and there is no radium to make good its loss. In 3.8 days about one half the emanation is transmuted and the Radium C decreases proportionately and, of course, with the Radium C the valuable radiations also decrease. In another 3.8 days—that is in about a week from the start—the radioactive value of the tube has fallen to one-fourth of its original value.

But in spite of the inconstant character of the emanation tube there are many reasons for preferring its use to the use of the radium tube. Chief of these is the fact that we can keep the precious radium safely locked up in the laboratory and not exposed to the thousand-and-one risks of the hospital. Then, secondly, the emanation, being a gas, is very convenient for subdivision into a large number of very small tubes according to the dosage required.

In fact the volume of the emanation is exceedingly minute. The amount of emanation in equilibrium with one gramme of radium is called the curie, and with one

milligramme the millicurie. Now, the volume of the curie is only a little more than one half a cubic millimetre. Hence in dealing with emanation from twenty or forty milligrammes of radium we are dealing with very small volumes.

How may the emanation be obtained? The process is an easy one in skilled and practised hands. The salt of radium—generally the bromide or chloride—is brought into acid solution. This causes the emanation to be freely given off as fast as it is formed. At intervals we pump it off with a mercury pump.

Let us see how many millicuries we will in future be able to turn out in the week in our new Dublin Radium Institute.[1] We shall have about 130 milligrammes of radium. In 3.8 days we get 65 millicuries from this—_i.e._ half the equilibrium amount of 130 millicuries. Hence in the week, we shall have about 130 millicuries.

This is not much. Many experts consider this little enough for one tube. But here in Dublin we have been using the emanation in a more economical and effective manner than is the usage elsewhere; according to a method which has been worked out and developed in our own Radium Institute. The economy is obtained by the very simple expedient of minutely subdividing the' dose. The system in vogue, generally, is to treat the tumour by inserting into it one or two very active

[1] Then recently established by the Royal Dublin Society.

tubes, containing, perhaps, up to 200 millicuries, or even more, per tube. Now these very heavily charged tubes give a radiation so intense at points close to the tube, due to the greater density of the rays near the tube, and, also, to the action of the softer and more easily absorbable rays, that it has been found necessary to stop these softer rays—both the y and ß—by wrapping lead or platinum round the tube. In this lead or platinum some thirty per cent. or more of the rays is absorbed and, of course, wasted. But in the absence of the screen there is extensive necrosis of the tissues near the tubes.

If, however, in place of one or two such tubes we use ten or twenty, each containing one-tenth or one-twentieth of the dose, we can avail ourselves of the softer rays around each tube with benefit. Thus a wasteful loss is avoided. Moreover a more uniform "illumination" of the tissues results, just as we can illuminate a hall more uniformly by the use of many lesser centres of light than by the use of one intense centre of radiation. Also we get what is called "cross-radiation,"which is found to be beneficial. The surgeon knows far better what he is doing by this method. Thus it may be arranged for the effects to go on with approximate uniformity throughout the tumour instead of varying rapidly around a central point or—and this may be very important— the effects may be readily concentrated locally.

Finally, not the least of the benefit arises in the easy technique of this new method. The quantities of

emanation employed can fit in the finest capillary glass tubing and the hairlike tubes can in turn be placed in fine exploring needles. There is comparatively little inconvenience to the patient in inserting these needles, and there is the most perfect control of the dosage in the number and strength of these tubes and the duration of exposure.[1]

The first Radium Institute in Ireland has already done good work for the relief of human suffering. It will have, I hope, a great future before it, for I venture, with diffidence, to hold the opinion, that with increased study the applications and claims of radioactive treatment will increase.

[1] For particulars of the new technique and of some of the work already accomplished, see papers, by Dr. Walter C. Stevenson, _British Medical Journal_, July 4th, 1914, and March 20th, 1915.

SKATING [1]

IT is now many years ago since, as a student, I was present at a college lecture delivered by a certain learned professor on the subject of friction. At this lecture a discussion arose out of a question addressed to our teacher: "How is it we can skate on ice and on no other substance?"

The answer came back without hesitation: "Because the ice is so smooth."

It was at once objected: "But you can skate on ice which is not smooth."

This put the professor in a difficulty. Obviously it is not on account of the smoothness of the ice. A piece of polished plate glass is far smoother than a surface of ice after the latter is cut up by a day's skating. Nevertheless, on the scratched and torn ice-surface skating is still quite possible; on the smooth plate glass we know we could not skate.

Some little time after this discussion, the connection between skating and a somewhat abstruse fact in physical science occurred to me. As the fact itself is one which has played a part in the geological history of the earth,

[1] A lecture delivered before the Royal Dublin Society in 1905.

and a part of no little importance, the subject of skating, whereby it is perhaps best brought home to every one, is deserving of our careful attention. Let not, then, the title of this lecture mislead the reader as to the importance of its subject matter.

Before going on to the explanation of the wonderful freedom of the skater's movements, I wish to verify what I have inferred as to the great difference in the slipperiness of glass and the slipperiness of ice. Here is a slab of polished glass. I can raise it to any angle I please so that at length this brass weight of 250 grams just slips down when started with a slight shove. The angle is, as you see, about 12½ degrees. I now transfer the weight on to this large slab of ice which I first rapidly dry with soft linen. Observe that the weight slips down the surface of ice at a much lower angle. It is a very low angle indeed: I read it as between 4 and 5 degrees. We see by this experiment that there is a great difference between the slipperiness of the two surfaces as measured by what is called "the angle of friction." In this experiment, too, the glass possesses by far the smoother surface although I have rubbed the deeper rugosities out of the ice by smoothing it with a glass surface. Notwithstanding this, its surface is spotted with small cavities due to bubbles and imperfections. It is certain that if the glass was equally rough, its angle of friction towards the brass weight would be higher.

We have, however, another comparative experiment to carry out. I made as you saw a determination of the angle at which this weight of 250 grams just slipped on the ice. The lower surface of the weight, the part which presses on the ice, consists of a light, brass curtain ring. This can be detached. Its mass is only 6½ grams, the curtain ring being, in fact, hollow and made of very thin metal. We have, therefore, in it a very small weight which presents exactly the same surface beneath as did the weight of 250 grams. You see, now, that this light weight will not slip on ice at 5 or 6 degrees of slope, but first does so at about io degrees.

This is a very important experiment as regards our present inquiry. Ice appears to possess more than one angle of friction according as a heavy or a light weight is used to press upon it. We will make the same experiment with the plate of glass. You see that there is little or no difference in the angle of friction of brass on glass when we press the surfaces together with a heavy or with a light weight. The light weight requires the same slope of 12½ degrees to make it slip.

This last result is in accordance with the laws of friction. We say that when solid presses on solid, for each pair of substances pressed together there is a constant ratio between the force required to keep one in motion over the other, and the force pressing the solids together. This ratio is called"the coefficient of friction."The coefficient is, in fact, constant or approximately so. I can determine the coefficient of friction from the angle of friction by taking the tangent of the angle. The tangent of the angle of friction is the coefficient of friction. If, then, the coefficient is constant, so, of course, must the angle of friction be constant. We have seen that it is so in the case of metal on glass, but not so in the case of metal on ice. This curious result shows that there is something abnormal about the slipperiness of ice.

The experiments we have hitherto made are open to the reproach that the surface of the ice is probably damp owing to the warmth of the air in contact with it. I have here a means of dealing with a surface of cold, dry ice. This shallow copper tank about 18 inches (45 cms.) long, and 4 inches (10 cms.) wide, is filled with a freezing 'mixture circulated through it from a larger vessel containing ice melting in hydrochloric acid at a temperature of about -18° C. This keeps the tank below the melting point of ice. The upper surface of the tank is provided with raised edges so that it can be flooded with water. The water is now frozen and its temperature is below 0° C. It is about 10° C. I can place over the ice a roof-shaped cover made of two inclined slabs of thick plate glass. This acts to keep out warm air, and to do away with any possibility of the surface of the ice being wet with water thawed from the ice. The whole tank along with its roof of glass can be adjusted to any angle, and a, scale at the

raised end of the tank gives the angle of slope in degrees. A weight placed on the ice can be easily seen through the glass cover.

The weight we shall use consists of a very light ring of aluminium wire which is rendered plainly visible by a ping-pong ball attached above it. The weight rests now on a copper plate provided for the purpose at the upper end of the tank. The plate being in direct contact beneath with the freezing mixture we are sure that the aluminium ring is no hotter than the ice. A light jerk suffices to shake the weight on to the surface of the ice.

We find that this ring loaded with only the ping-pong ball, and weighing a total of 2.55 grams does not slip at the low angles. I have the surface of the ice at an angle of rather over 13½, and only by continuous tapping of the apparatus can it be induced to slip down. This is a coefficient of 0.24, and compares with the coefficient of hard and smooth solids on one another. I now replace the empty ping-pong ball by a similar ball filled with lead shot. The total weight is now 155 grams. You see the angle of slipping has fallen to 7°.

Every one who has made friction experiments knows how unsatisfactory and inconsistent they often are. We can only discuss notable quantities and broad results, unless the most conscientious care be taken to eliminate errors. The net result here is that ice at about -10° C. when pressed on by a very light weight possesses a

coefficient of friction comparable with the usual coefficients of solids on solids, but when the pressure is increased, the coefficient falls to about half this value.

The following table embodies some results obtained on the friction of ice and glass, using the methods I have shown you. I add some of the more carefully determined coefficients of other observers.

	Wt. in Grams.	On Plate Glass.		On Ice at 0° C.		On Ice at 10° C.	
		Angle.	Coeff.	Angle.	Coeff.	Angle.	Coeff
Aluminium	2.55	12½°	0.22	12°	0.21	13½°	0.24
Same	155	12½°	0.22	6°	0.11	7°	0.12
Brass	6.5	12½°	0.22	10°	0.17	10½°	0.18
Same	107	12½°	0.22	5°	0.09	6°	0.10

Steel on steel (Morin) - - - - 0.14
Brass on cast iron (Morin) - - 0.19
Steel on cast iron (Morin) - - 0.20
Skate on ice (J. Müller) - - - 0.016—0.032
Best-greased surfaces (Perry) - 0.03—0.036

You perceive from the table that while the friction of brass or aluminium on glass is quite independent of the weight used, that of brass or aluminium on ice depends in some way upon the weight, and falls in a very marked degree when the weight is heavy. Now, I think that if we had been on the look out for any abnormality in the friction of hard substances on ice, we would have rather anticipated a variation in the other direction. We would have, perhaps, expected that a heavy weight would have given rise to the greater friction. I now turn to the explanation of this extraordinary result.

You are aware that it requires an expenditure of heat merely to convert ice to water, the water produced being at the temperature of the ice, _i.e._ at 0° C., from which it is derived. The heat required to change the ice from the solid to the liquid state is the latent heat of water. We take the unit quantity of heat to be that which is required to heat 1 kilogram of water 1° C. Then if we melt 1 kilogram of ice, we must supply it with 80 such units of heat. While melting is going on, there is no change of temperature if the experiment is carefully conducted. The melting ice and the water coming from it remain at 0° C. throughout the operation, and neither the thermometer nor your own sensations would tell you of the amount of heat which was flowing in. The heat is latent or hidden in the liquid produced, and has gone to do molecular work in the substance. Observe that if we supply only 40 thermal units, we get only one-half the ice melted. If only 10 units are supplied, then we get only one eighth of a kilogram of water, and no more nor less.

I have ventured to recall to you these commonplaces of science before considering a mode of melting ice which is less generally known, and which involves no supply of heat on your part. This method involves for its

understanding a careful consideration of the thermal properties of water in the solid state.

It must have been observed a very long time ago that water expands when it freezes. Otherwise ice would not float on water; and, what is perhaps more important in your eyes, your water pipes would not burst in winter when the water freezes therein. But although the important fact of the expansion of water on freezing was so long presented to the observation of mankind, it was not till almost exactly the middle of the last century that James Thomson, a gifted Irishman, predicted many important consequences arising from the fact of the expansion of water on becoming solid. The principles lie enunciated are perfectly general, and apply in every case of change of volume attending change of state. We are here only concerned with the case of water and ice.

James Thomson, following a train of thought which we cannot here pursue, predicted that owing to the fact of the expansion of water on becoming solid, pressure will lower the melting point of ice or the freezing point of water. Normally, as you are aware, the temperature is 0° C. or 32° F. Thomson said that this would be found to be the freezing point only at atmospheric pressure. He calculated how much it would change with change of pressure. He predicted that the freezing point would fall 0.0075 of a degree Centigrade for each additional atmosphere of pressure applied to the water. Suppose,

for instance, our earth possessed an atmosphere so heavy to as exert a thousand times the pressure of the existing atmosphere, then water would not freeze at 0° C., but at -7.5° C. or about 18° F. Again, in vacuo, that is when the pressure has been reduced to the relatively small vapour pressure of the water, the freezing point is above 0° C., _i.e._ at 0.0075° C. In parts of the ocean depths the pressure is much over a thousand atmospheres. Fresh water would remain liquid there at temperatures much below 0° C.

It will be evident enough, even to those not possessed of the scientific insight of James Thomson, that some such fact is to be anticipated. It is, however, easy to be wise after the event. It appeals to us in a general way that as water expands on freezing, pressure will tend to resist the turning of it to ice. The water will try to remain liquid in obedience to the pressure. It will, therefore, require a lower temperature to induce it to become ice.

James Thomson left his thesis as a prediction. But he predicted exactly what his distinguished brother, Sir William Thomson—later Lord Kelvin—found to happen when the matter was put to the test of experiment. We must consider the experiment made by Lord Kelvin.

According to Thomson's views, if a quantity of ice and water are compressed, there must be _a fall of temperature_. The nature of his argument is as follows:

Let the ice and water be exactly at 0° C. to start with. Then suppose we apply, say, one thousand atmospheres pressure. The melting point of the ice is lowered to -7.5° C. That is, it will require a temperature so low as -7.5° C. to keep it solid. It will therefore at once set about melting, for as we have seen, its actual temperature is not -7.5° C., but a higher temperature, _i.e._ 0° C. In other words, it is 7.5° above its melting point. But as soon as it begins melting it also begins to absorb heat to supply the 80 thermal units which, as we know, are required to turn each kilogram of the ice to water. Where can it get this heat? We assume that we give it none. It has only two sources, the ice can take heat from itself, and it can take heat from the water. It does both in this case, and both ice and water drop in temperature. They fall in temperature till -7.5° is reached. Then the ice has got to its melting point under the pressure of one thousand atmospheres, or, as we may put it, the water has reached its freezing point. There can be no more melting. The whole mass is down to -7.5° C., and will stay there if we keep heat from flowing either into or out of the vessel. There is now more water and less ice in the vessel than when we started, and the temperature has fallen to -7.5° C. The fall of temperature to the amount predicted by the theory was verified by Lord Kelvin.

Suppose we now suddenly remove the pressure; what will happen? We have water and ice at -7.5° C.

and at the normal pressure. Water at -7.5° and at the normal pressure of course turns to ice. The water will, therefore, instantly freeze in the vessel, and the whole process will be reversed. In freezing, the water will give up its latent heat, and this will warm up the whole mass till once again 0° C. is attained. Then there will be no more freezing, for again the ice is at its melting point. This is the remarkable series of events which James Thomson predicted. And these are the events which Lord Kelvin by a delicate series of experiments, verified in every respect.

Suppose we had nothing but solid ice in the vessel at starting, would the experiment result in the same way? Yes, it assuredly would. The ice under the increased pressure would melt a little everywhere throughout its mass, taking the requisite latent heat from itself at the expense of its sensible heat, and the temperature of the ice would fall to the new melting point.

Could we melt the whole of the ice in this manner? Again the answer is "yes." But the pressure must be very great. If we assume that all the heat is obtained at the expense of the sensible heat of the ice, the cooling must be such as to supply the latent heat of the whole mass of water produced. However, the latent heat diminishes as the melting point is lowered, and at a rate which would reduce it to nothing at about 18,000 atmospheres. Mousson, operating on ice enclosed in a conducting cylinder and cooled to -18° at starting

appears to have obtained very complete liquefaction. Mousson must have attained a pressure of at least an amount adequate to lower the melting point below -18°. The degree of liquefaction actually attained may have been due in part to the passage of heat through the walls of the vessel. He proved the more or less complete liquefaction of the ice within the vessel by the fall of a copper index from the top to the bottom of the vessel while the pressure was on.

I have here a simple way of demonstrating to you the fall of temperature attending the compression of ice. In this mould, which is strongly made of steel, lined with boxwood to diminish the passage of conducted heat, is a quantity of ice which I compress when I force in this plunger. In the ice is a thermoelectric junction, the wires leading to which are in communication with a reflecting galvanometer. The thermocouple is of copper and nickel, and is of such sensitiveness as to show by motion of the spot of light on the screen even a small fraction of a degree. On applying the pressure, you see the spot of light is displaced, and in such a direction as to indicate cooling. The balancing thermocouple is all the time imbedded in a block of ice so that its temperature remains unaltered. On taking off the pressure, the spot of light returns to its first position. I can move the spot of light backwards and forwards on the screen by taking off and putting on the pressure. The effects are quite instantaneous.

The fact last referred to is very important. The ice, in fact, is as it were automatically turned to water. It is not a matter of the conduction of heat from point to point in the ice. Its own sensible heat is immediately absorbed throughout the mass. This would be the theoretical result, but it is probable that owing to imperfections throughout the ice and failure in uniformity in the distribution of the stress, the melting would not take place quite uniformly or homogeneously.

Before applying our new ideas to skating, I want you to notice a fact which I have inferentially stated, but not specifically mentioned. Pressure will only lead to the melting of ice if the new melting point, _i.e._ that due to the pressure, is below the prevailing temperature. Let us take figures. The ice to start with is, say, at -3° C. Suppose we apply such a pressure to this ice as will confer a melting point of -2° C. on it. Obviously, there will be no melting. For why should ice which is at -3° C. melt when its melting point is -2° C.? The ice is, in fact, colder than its melting point. Hence, you note this fact: The pressure must be sufficiently intense to bring the melting point below the prevailing temperature, or there will be no melting; and the further we reduce the melting point by pressure below the prevailing temperature, the more ice will be melted.

We come at length to the object of our remarks I don't know who invented skating or skates. It is said that in the thirteenth century the inhabitants of

England used to amuse themselves by fastening the bones of an animal beneath their feet, and pushing themselves about on the ice by means of a stick pointed with iron. With such skates, any performance either on inside or outside edge was impossible. We are a conservative people. This exhilarating amusement appears to have served the people of England for three centuries. Not till 1660 were wooden skates shod with iron introduced from the Netherlands. It is certain that skating was a fashionable amusement in Pepys' time. He writes in 1662 to the effect: "It being a great frost, did see people sliding with their skates, which is a very pretty art." It is remarkable that it was the German poet Klopstock who made skating fashionable in Germany. Until his time, the art was considered a pastime, only fit for very young or silly people.

I wish now to dwell upon that beautiful contrivance the modern skate. It is a remarkable example of how an appliance can develop towards perfection in the absence of a really intelligent understanding of the principles underlying its development. For what are the principles underlying the proper construction of the skate? After what I have said, I think you will readily understand. The object is to produce such a pressure under the blade that the ice will melt. We wish to establish such a pressure under the skate that even on a day when the ice is below zero, its melting

point is so reduced just under the edge of the skate that the ice turns to water.

It is this melting of the ice under the skate which secures the condition essential to skating. In the first place, the skate no longer rests on a solid. It rests on a liquid. You are aware how in cases where we want to reduce friction—say at the bearing of a wheel or under a pivot—we introduce a liquid. Look at the bearings of a steam engine. A continuous stream of oil is fed in to interpose itself between the solid surfaces. I need not illustrate so well-known a principle by experiment. Solid friction disappears when the liquid intervenes. In its place we substitute the lesser difficulty of shearing one layer of the liquid over the other; and if we keep up the supply of oil the work required to do this is not very different, no matter how great we make the pressure upon the bearings. Compared with the resistance of solid friction, the resistance of fluid friction is trifling. Here under the skate the lubrication is perhaps the most perfect which it is possible to conceive. J. Müller has determined the coefficient by towing a skater holding on by a spring balance. The coefficient is between 0.016 and 0.032. In other words, the skater would run down an incline so little as 1 or 2 degrees; an inclination not perceivable by the eye. Now observe that the larger of these coefficients is almost exactly the same as that which Perry found in the case of well-greased surfaces. But evidently no

artificial system of lubrication could hope to equal that which exists between the skate and the ice. For the lubrication here is, as it were, automatic. In the machine if the lubricant gets squeezed out there instantly ensues solid friction. Under the skate this cannot happen for the squeezing out of the lubricant is instantly followed by the formation of another film of water. The conditions of pressure which may lead to solid friction in the machine here automatically call the lubricant into existence.

Just under the edge of the skate the pressure is enormous. Consider that the whole weight of the skater is born upon a mere knife edge. The skater alternately throws his whole weight upon the edge of each skate. But not only is the weight thus concentrated upon one edge, further concentration is secured in the best skates by making the skate hollow-ground, _i.e._ increasing the keenness of the edge by making it less than a right angle. Still greater pressure is obtained by diminishing the length of that part of the blade which is in contact with the ice. This is done by putting curvature on the blade or making it what is called "hog-backed." You see that everything is done to diminish the area in contact with the ice, and thus to increase the pressure. The result is a very great compression of the ice beneath the edge of the skate. Even in the very coldest weather melting must take place to some extent.

As we observed before, the melting is instantaneous,

Heat has not to travel from one point of the ice to another; immediately the pressure comes on the ice it turns to water. It takes the requisite heat from itself in order that the change of state may be accomplished. So soon as the skate passes on, the water resumes the solid state. It is probable that there is an instantaneous escape, and re-freezing of some of the water from beneath the skate, the skate instantly taking a fresh bearing and melting more ice. The temperature of the water escaping from beneath the skate, or left behind by it, immediately becomes what it was before the skate pressed upon it.

Thus, a most wonderful and complex series of molecular events takes place beneath the skate. Swift as it passes, the whole sequence of events which James Thomson predicted has to take place beneath the blade Compression; lowering of the melting point below the temperature of the surrounding ice; melting; absorption of heat; and cooling to the new melting point, _i.e._ to that proper to the pressure beneath the blade. The skate now passes on. Then follow: Relief of pressure; re-solidification of the water; restoration of the borrowed heat from the congealing water and reversion of the ice to the original temperature.

If we reflect for a moment on all this, we see that we do not skate on ice but on water. We could not skate on ice any more than we could skate on glass. We saw that with light weights and when the pressure

{Diagram}

Diagram showing successive states obtaining in ice, before, during, and after the passage of the skate. The temperatures and pressures selected for illustration are such as might occur under ordinary conditions. The edge of the skate is shown in magnified cross-section.

Was not sufficient to melt the ice, the friction was much the same as that of metal on glass. Ice is not slippery. It is an error to say that it is. The learned professor was very much astray when he said that you could skate on ice because it is so smooth. The smoothness of the ice has nothing to do with the matter. In short, owing to the action of gravity upon your body, you escape the normal resistance of solid on solid, and glide about with feet winged like the messenger of the Gods; but on water.

A second condition essential to the art of skating is also involved in the melting of the ice. The sinking of the skate gives the skater "bite." This it is which enables him to urge himself forward. So long as skates consisted of the rounded bones of animals, the skater had to use a pointed staff to propel himself. In creating bite, the skater again unconsciously appeals to the peculiar physical properties of ice. The pressure required for the propulsion of the skater is spread all along the length of the groove he has cut in the ice, and obliquely downwards. The skate will not slip away laterally, for the horizontal component of the pressure is not enough to melt the ice. He thus gets the resistance he requires.

You see what a very perfect contrivance the skate is; and what a similitude of intelligence there is in its evolution. Blind intelligence, because it is certain the true physics of skating was never held in view by

the makers of skates. The evolution of the skate has been truly organic. The skater selected the fittest skate, and hence the fit skate survived.

In a word, the possibility of skating depends on the dynamical melting of ice under pressure. And observe the whole matter turns upon the apparently unrelated fact that the freezing of water results in a solid more bulky than the water which gives rise to it. If ice was less bulky than the water from which it was derived, pressure would not melt it; it would be all the more solid for the pressure, as it were. The melting point would rise instead of falling. Most substances behave in this manner, and hence we cannot skate upon them. Only quite a few substances expand on freezing, and it happens that their particular melting temperatures or other properties render them unsuitable to skating. The most abundant fluid substance on the earth, and the most abundant substance of any one kind on its surface, thus possesses the ideally correct and suitable properties for the art of skating.

I have pointed out that the pressure must be such as to bring the temperature of melting below that prevailing in the ice at the time. We have seen also, that one atmosphere lowers the melting point of ice by the 1/140 of a degree Centigrade; more exactly by 0.0075°. Let us now assume that the skate is so far sunken in the ice as to bear for a length of two inches, and for a width of one-hundredth of an inch. The skater weighs,

let us say—150 pounds. If this weight was borne on one square inch, the pressure would be ten atmospheres. But the skater rests his weight, in fact, upon an area of one-fiftieth of an inch. The pressure is, therefore, fifty times as great. The ice is subjected to a pressure of 500 atmospheres. This lowers the melting point to -3.75° C. Hence, on a day when the ice is at this temperature, the skate will sink in the ice till the weight of the skater is concentrated as we have assumed. His skate can sink no further, for any lesser concentration of the pressure will not bring the melting point below the prevailing temperature. We can calculate the theoretical bite for any state of the ice. If the ice is colder the bite will not be so deep. If the temperature was twice as far below zero, then the area over which the skater's weight will be distributed, when the skate has penetrated its maximum depth, will be only half the former area, and the pressure will be one thousand atmospheres.

An important consideration arises from the fact that under the very extreme edge of the skate the pressure is indefinitely great. For this involves that there will always be some bite, however cold the ice may be. That is, the narrow strip of ice which first receives the skater's weight must partially liquefy however cold the ice.

It must have happened to many here to be on ice which was too cold to skate on with comfort. The

skater in this case speaks of the ice as too hard. In the Engadine, the ice on the large lakes gets so cold that skaters complain of this. On the rinks, which are chiefly used there, the ice is frequently renewed by flooding with water at the close of the day. It thus never gets so very cold as on the lakes. I have been on ice in North France, which, in the early morning, was too hard to afford sufficient bite for comfort. The cause of this is easily understood from what we have been considering.

We may now return to the experimental results which we obtained early in the lecture. The heavy weights slip off the ice at a low angle because just at the points of contact with the ice the latter melts, and they, in fact, slip not on ice but on water. The light weights on cold, dry ice do not lower the melting point below the temperature of the ice, _i.e._ below -10° C., and so they slip on dry ice. They therefore give us the true coefficient of friction of metal on ice.

This subject has, more recently been investigated by H. Morphy, of Trinity College, Dublin. The refinement of a closed vessel at uniform temperature, in which the ice is formed and the experiment carried out, is introduced. Thermocouples give the temperatures, not only of the ice but of the aluminium sleigh which slips upon it under various loads. In this way we may be certain that the metal runners are truly at the temperature of the ice. I now quote from Morphy's paper

"The angle of friction was found to remain constant until a certain stage of the loading, when it suddenly fell to about half of its original value. It then remained constant for further increases in the load.

"These results, which confirmed those obtained previously with less satisfactory apparatus, are shown in the table below. In the first column is shown the load, _i.e._ the weight of sleigh + weight of shot added. In the second and third columns are shown, respectively, the coefficient and angle of friction, whilst the fourth gives the temperature of the ice as determined from the galvanometer deflexions.

Load.	Tan y.	y.	Temp.
5.68 grams.	0.36±.01	20°±30'	-5.65° C.
10.39			-5.65°
11.96			-5.75°
12.74			-5.60°
13.53			-5.65°
14.31			-5.65°
15.10 grams.	0.17±.01	9°.30'±30'	-5.60°
16.67			-5.55°
19.81			-5.60°
24.52			-5.60°
5.68 grams.	0.36±.01	20°±30'	-5.60°

"These experiments were repeated on another occasion with the same result and similar results had been obtained with different apparatus.

"As a result of the investigation the following points are clearly shown:—

"(1) The coefficient of friction for ice at constant temperature may have either of two constant values according to the pressure per unit surface of contact.

"(2) For small pressures, and up to a certain well defined limit of pressure, the coefficient is fairly large, having the value 0.36±.01 in the case investigated.

"(3) For pressures greater than the above limit the coefficient is relatively small, having the value 0.17±.01 in the case investigated."

It will be seen that Morphy's results are similar to those arrived at in the first experimental consideration of our subject; but from the manner in which the experiments have been carried out, they are more accurate and reliable.

A great deal more might be said about skating, and the allied sports of tobogganing, sleighing, curling, ice yachting, and last, but by no means least, sliding—that unpretentious pastime of the million. Happy the boy who has nails in his boots when Jack-Frost appears in his white garment, and congeals the neighbouring pond. But I must turn away at the threshold of the humorous aspect of my subject (for the victim of the street "slide" owes his injured dignity to the abstruse laws we have been discussing) and pass to other and graver subjects intimately connected with skating.

James Thomson pointed out that if we apply compressional stress to an ice crystal contained in a vessel

which also contains other ice crystals, and water at 0° C., then the stressed crystal will melt and become water, but its counterpart or equivalent quantity of ice will reappear elsewhere in the vessel. This is, obviously, but a deduction from the principles we have been examining. The phenomenon is commonly called "regelation." I have already made the usual regelation experiment before you when I compressed broken ice in this mould. The result was a clear, hard and almost flawless lens of ice. Now in this operation we must figure to ourselves the pieces of ice when pressed against one another melting away where compressed, and the water produced escaping into the spaces between the fragments, and there solidifying in virtue of its temperature being below the freezing point of unstressed water. The final result is the uniform lens of ice. The same process goes on in a less perfect manner when you make—or shall I better say—when you made snowballs.

We now come to theories of glacier motion; of which there are two. The one refers it mainly to regelation; the other to a real viscosity of the ice.

The late J. C. M'Connel established the fact that ice possesses viscosity; that is, it will slowly yield and change its shape under long continued stresses. His observations, indeed, raise a difficulty in applying this viscosity to explain glacier motion, for he showed that an ice crystal is only viscous in a certain structural

direction. A complex mixture of crystals such, as we know glacier ice to be, ought, we would imagine, to display a nett or resultant rigidity. A mass of glacier ice when distorted by application of a force must, however, undergo precisely the transformations which took place in forming the lens from the fragments of ice. In fact, regelation will confer upon it all the appearance of viscosity.

Let us picture to ourselves a glacier pressing its enormous mass down a Swiss valley. At any point suppose it to be hindered in its downward path by a rocky obstacle. At that point the ice turns to water just as it does beneath the skate. The cold water escapes and solidifies elsewhere. But note this, only where there is freedom from pressure. In escaping, it carries away its latent heat of liquefaction, and this we must assume, is lost to the region of ice lately under pressure. This region will, however, again warm up by conduction of heat from the surrounding ice, or by the circulation of water from the suxface. Meanwhile, the pressure at that point has been relieved. The mechanical resistance is transferred elsewhere. At this new point there is again melting and relief of pressure. In this manner the glacier may be supposed to move down. There is continual flux of conducted heat and converted latent heat, hither and thither, to and from the points of resistance. The final motion of the whole mass is necessarily slow; a few feet in the day or, in winter,

even only a few inches. And as we might expect, perfect silence attends the downward slipping of the gigantic mass. The motion is, I believe, sufficiently explained as a skating motion. The skate is, however, fixed, the ice moves. The great Aletsch Glacier collects its snows among the highest summits of the Oberland. Thence, the consolidated ice makes its way into the Rhone Valley, travelling a distance of some 20 miles. The ice now melting into the youthful Rhone fell upon the Monch, the Jungfrau or the Eiger in the days when Elizabeth ruled in England and Shakespeare lived.

The ice-fall is a common sight on the glacier. In great lumps and broken pinnacles it topples over some rocky obstacle and falls shattered on to the glacier below. But a little further down the wound is healed again, and regelation has restored the smooth surface of the glacier. All such phenomena are explained on James Thomson's exposition of the behaviour of a substance which expands on passing from the liquid to the solid state.

We thus have arrived at very far-reaching considerations arising out of skating and its science. The tendency for snow to accumulate on the highest regions of the Earth depends on principles which we cannot stop to consider. We know it collects above a certain level even at the Equator. We may consider, then, that but for the operation of the laws which James Thomson brought to light, and which his illustrious brother,

Lord Kelvin, made manifest, the uplands of the Earth could not have freed themselves of the burthen of ice. The geological history of the Earth must have been profoundly modified. The higher levels must have been depressed; the general level of the ocean relatively to the land thereby raised, and, it is even possible, that such a mean level might have been attained as would result in general submergence.

During the last great glacial period, we may say the fate of the world hung on the operation of those laws which have concerned us throughout this lecture. It is believed the ice was piled up to a height of some 6,000 feet over the region of Scandinavia. Under the influence of the pressure and fusion at points of resistance, the accumulation was stayed, and it flowed southwards the accumulation was stayed, and it flowed southwards over Northern Europe. The Highlands of Scotland were covered with, perhaps, three or four thousand feet of ice. Ireland was covered from north to south, and mighty ice-bergs floated from our western and southern shores.

The transported or erratic stones, often of great size, which are found in many parts of Ireland, are records of these long past events: events which happened before Man, as a rational being, appeared upon the Earth.

A SPECULATION AS TO A PREMATERIAL UNIVERSE [1]

"And therefore...these things likewise had a birth; for things which are of mortal body could not for an infinite time back... have been able to set at naught the puissant strength of immeasurable age."—LUCRETIUS, _De Rerum Natura._

"O fearful meditation! Where, alack! Shall Time's best jewel from Time's chest lie hid?" —SHAKESPEARE.

IN the material universe we find presented to our senses a physical development continually progressing, extending to all, even the most minute, material configurations. Some fundamental distinctions existing between this development as apparent in the organic and the inorganic systems of the present day are referred to elsewhere in this volume.[2] In the present essay, these systems as having a common origin and common ending, are merged in the same consideration as to the nature of the origin of material systems in general. This present essay is occupied by the consideration of the necessity of limiting material interactions in past time. The speculation originated in the difficulties which present themselves when we ascribe to these interactions infinite duration in the past. These difficulties first claim our consideration.

[1] Proc. Royal Dublin Soc., vol. vii., Part V, 1892.

[2] _The Abundance of Life._

Accepting the hypothesis of Kant and Laplace in its widest extension, we are referred to a primitive condition of wide material diffusion, and necessarily too of material instability. The hypothesis is, in fact, based upon this material instability. We may pursue the sequence of events assumed in this hypothesis into the future, and into the past.

In the future we find finality to progress clearly indicated. The hypothesis points to a time when there will be no more progressive change but a mere sequence of unfruitful events, such as the eternal uniform motion of a mass of matter no longer gaining or losing heat in an ether possessed of a uniform distribution of energy in all its parts. Or, again, if the ether absorb the energy of material motion, this vast and dark aggregation eternally poised and at rest within it. The action is transferred to the subtle parts of the ether which suffer none of the energy to degrade. This is, physically, a thinkable future. Our minds suggest no change, and demand none. More than this, change is unthinkable according to our present ideas of energy. Of progress there is an end.

This finality _â parte post_ is instructive. Abstract considerations, based on geometrical or analytical illustrations, question the finiteness of some physical developments. Thus our sun may require eternal time to attain the temperature of the ether around it, the approach to this condition being assumed to be asymptotic in

character. But consider the legitimate _reductio ad absurdum_ of an ember raked from a fire 1000 years ago. Is it not yet cooled down to the constant temperature of its surroundings? And we may evidently increase the time a million-fold if we please. It appears as if we must regard eternity as outliving every progressive change, For there is no convergence or enfeeblement of time. The ever-flowing present moves no differently for the occurrence of the mightiest or the most insignificant events. And even if we say that time is only the attendant upon events, yet this attendant waits patiently for the end, however long deferred.

Does the essentially material hypothesis of Kant and Laplace account for an infinite past as thinkably as it accounts for the infinite future? As this hypothesis is based upon material instability the question resolves itself into this:— Is the assumption of an infinitely prolonged past instability a probable or possible account of the past? There are, it appears to me, great difficulties involved in accepting the hypothesis of infinitely prolonged material instability. I will refer here to three principal objections. The first may be called a metaphysical objection; the second is partly metaphysical and partly physical, the third may be considered a physical objection, as it is involved directly in the phenomena presented by our universe.

The metaphysical objection must have presented itself to every one who has considered the question. It may be put thus:—If present events are merely one stage in an infinite progress, why is not the present stage long ago passed over? We are evidently at liberty to push back any stage of progress to as remote a period as we like by putting back first the one before this and next the stage preceding this, and so on, for, by hypothesis, there is no beginning to the progress.

Thus, the sum of passing events constituting the present universe should long ago have been accomplished and passed away. If we consider alternative hypotheses not involving this difficulty, we are at once struck by the fact that the future of material development is free of the objection. For the eternity of unprogressive events involved in the future on Kant's hypothesis, is not only thinkable, but any change is, as observed, irreconcilable with our ideas of energy. As in the future so in the past we look to a cessation to progress. But as we believe the activity of the present universe must in some form have existed all along, the only refuge in the past is to imagine an active but unprogressive eternity, the unprogressive activity at some period becoming a progressive activity—that progressive activity of which we are spectators. To the unprogressive activity there was no beginning; in fact, beginning is as unthinkable and uncalled for to the unprogressive activity of the past as ending is to the unprogressive activity of the future, when all developmental actions shall have ceased. There is no beginning or ending to the activity of the universe.

There is beginning and ending to present progressive activity. Looking through the realm of nature we seek beginning and ending, but "passing through nature to eternity" we find neither. Both are justified; the questioning of the ancient poet regarding the past, and of the modern regarding the future, quoted at the head of this essay.

The next objection, which is in part metaphysical, is founded on the difficulty of ascribing any ultimate reality or potency to forces diminishing through eternal time. Thus, against the assumption that our universe is the result of material aggregation progressing over eternal time, which involves the primitive infinite separation of the particles, we may ask, what force can have acted between particles sundered by infinite distance? The gravitational force falling off as the square of the distance, must vanish at infinity if we mean what we say when we ascribe infinite separation to them. Their condition is then one of neutral stability, a finite movement of the particles neither increasing nor diminishing interaction. They had then remained eternally in their separated condition, there being no cause to render such condition finite. The difficulty involved here appears to me of the same nature as the difficulty of ascribing any residual heat to the sun after eternal time has elapsed. In both cases we are bound to prolong the time, from our very idea of time, till progress is no more, when in the one case we can imagine no mutual approximation of the

particles, in the other no further cooling of the body. However, I will riot dwell further upon this objection, as it does not, I believe, present itself with equal force to every mind. A reason less open to dispute, as being less subjective, against the aggregation of infinitely remote particles as the origin of our universe, is contained in the physical objection.

In this objection we consider that the appearance presented by our universe negatives the hypothesis of infinitely prolonged aggregation. We base this negation upon the appearance of simultaneity ~ presented by the heavens, contending that this simultaneity is contrary to what we would expect to find in the case of particles gathered from infinitely remote distances. Whether these particles were endowed with relative motions or not is unimportant to the consideration. In what respects do the phenomena of our universe present the appearance of simultaneous phenomena? We must remember that the suns in space are as fires which brighten only for a moment and are then extinguished. It is in this sense we must regard the longest burning of the stars. Whether just lit or just expiring counts little in eternity. The light and heat of the star is being absorbed by the ether of space as effectually and rapidly as the ocean swallows the ripple from the wings of an expiring insect. Sir William Herschel says of the galaxy of the milky way:— "We do not know the rate of progress of this mysterious chronometer, but it is nevertheless certain that it cannot

last for ever, and its past duration cannot be infinite." We do not know, indeed, the rate of progress of the chronometer, but if the dial be one divided into eternal durations the consummation of any finite physical change represents such a movement of the hand as is accomplished in a single vibration of the balance wheel.

Hence we must regard the hosts of glittering stars as a conflagration that has been simultaneously lighted up in the heavens. The enormous (to our ideas) thermal energy of the stars resembles the scintillation of iron dust in a jar of oxygen when a pinch of the dust is thrown in. Although some particles be burnt up before others become alight, and some linger yet a little longer than the others, in our day's work the scintillation of the iron dust is the work of a single instant, and so in the long night of eternity the scintillation of the mightiest suns of space is over in a moment. A little longer, indeed, in duration than the life which stirs a moment in response to the diffusion of the energy, but only very little. So must an Eternal Being regard the scintillation of the stars and the periodic vibration of life in our geological time and the most enduring efforts of thought. The latter indeed are no more lasting than

"... the labour of ants In the light of a million million of suns."

But the myriad suns themselves, with their generations, are the momentary gleam of lights for ever after extinguished.

Again, science suggests that the present process of material aggregation is not finished, and possibly will only be when it prevails universally. Hence the very distribution of the stars, as we observe them, as isolated aggregations, indicates a development which in the infinite duration must be regarded as equally advanced in all parts of stellar space and essentially a simultaneous phenomenon. For were we spectators of a system in which any very great difference of age prevailed, this very great difference would be attended by some such appearance as the following:—

The aupearance of but one star, other generations being long extinct or no others yet come into being; or, perhaps, a faint nebulous wreath of aggregating matter somewhere solitary in the heavens; or no sign of matter beyond our system, either because ungathered or long passed away into darkness.[1]

Some such appearances were to be expected had the aggregation of matter depended solely on chance encounters of particles scattered through infinite space.

For as, by hypothesis, the aggregation occupies an infinite time in consummation it is nearly a certainty that each particle encountered after immeasurable time, and then for the first time endowed with actual gravitational potential energy, would have long expended this energy

[1] It is interesting to reflect upon the effect which an entire absence of luminaries outside our solar system would have had upon the views of our philosophers and upon our outlook on life.

before another particle was gathered. But the fact that so many fires which we know to be of brief duration are scattered through a region of space, and the fact of a configuration which we believe to be a transitory ore, suggest their simultaneous aggregation here and there. And in the nebulous wreaths situated amidst the stars there is evidence that these actually originated where they now are, for in such no relative motion, I believe, has as yet been detected by the spectroscope. All this, too, is in keeping with the nebular hypothesis of Kant and Laplace so long as this does not assume a primitive infinite dispersion of matter, but the gathering of matter from finite distances first into nebulous patches which aggregating with each other have given rise to our system of stars. But if we extend this hypothesis throughout an infinite past by the supposition of aggregation of infinitely remote particles we replace the simultaneous approach required in order to accotnt for the simultaneous phenomena visible in the heavens, by a succession of aggregative events, by hypothesis at intervals of nearly infinite duration, when the events of the universe had consisted of fitful gleams lighted after eternities of time and extinguished for yet other eternities.

Finally, if we seek to replace the eternal instability involved in Kant's hypothesis when extended over an infinite past, by any hypothesis of material stability, we at once find ourselves in the difficulty that from the known properties of matter such stability must have been permanent if ever existent, which is contrary to fact. Thus the kinetic inertia expressed in Newton's first law of motion might well be supposed to secure equilibrium with material attraction, but if primevally diffused matter had ever thus been held in equilibrium it must have remained so, or it was maintained so imperfectly, which brings us back to endless evolution.

On these grounds I contend that the present gravitational properties of matter cannot be supposed to have acted for all past duration. Universal equilibrium of gravitating particles would have been indestructible by internal causes. Perpetual instability or evolution is alike unthinkable and contrary to the phenomena of the universe of which we are cognisant. We therefore turn from gravitating matter as affording no rational account of the past. We do so of necessity, however much we feel our ignorance of the nature of the unknown actions to which we have recourse.

A prematerial condition of the universe was, we assume, a condition in which uniformity as regards the average distribution of energy in space prevailed, but neterogeneity and instability were possible. The realization of that possibility was the beginning we seek, and we today are witnesses of the train of events involved in the breakdown of an eternal past equilibrium. We are witnesses on this hypothesis, of a catastrophe possibly confined to certain regions of space, but which is, to the motions and configurations concerned, absolutely unique, reversible to

its former condition of potential by no process of which we can have any conception.

Our speculation is that we, as spectators of evolution, are witnessing the interaction of forces which have not always been acting. A prematerial state of the universe was one of unfruitful motions, that is, motions unattended by progressing changes, in our region of the ether. How extended we cannot say; the nature of the motions we know not; but the kinetic entities differed from matter in the one important particular of not possessing gravitational attraction. Such kinetic configurations we cannot consider to be matter. It was _possible_ to construct matter by their summation or linkage as the configuration of the crystal is possible in the clear supersaturated liquid.

Duration in an ether filled with such motions would pass in a succession of mere unfruitful events; as duration, we may imagine, even now passes in parts of the ether similar to our own. An endless (it may be) succession of unprogressive, fruitless events. But at one moment in the infinite duration the requisite configuration of the elementary motions is attained; solely by the one chance disposition the stability of all must go, spreading from the fateful point.

Possibly the material segregation was confined to one part of space, the elementary motions condensing upon transformation, and so impoverishing the ether around till the action ceased. Again in the same sense as the

stars are simultaneous, so also they may be regarded as uniform in size, for the difference in magnitude might have been anything we please to imagine, if at the same time we ascribe sufficient distance sundering great and small. So, too;, will a dilute solution of acetate of soda build a crystal at one point, and the impoverishment of the medium checking the growth in this region, another centre will begin at the furthest extremities of the first crystal till the liquid is filled with loose feathery aggregations comparable in size with one another. In a similar way the crystallizing out of matter may have given rise, not to a uniform nebula in space, but to detached nebula, approximately of equal mass, from which ultimately were formed the stars.

That an all-knowing Being might have foretold the ultimate event at any preceding period by observing the motions of the parts then occurring, and reasoning as to the train of consequences arising from these nations, is supposable. But considerations arising from this involve no difficulty in ascribing to this prematerial train of events infinite duration. For progress there is none, and we can quite as easily conceive of some part of space where the same Infinite Intelligence, contemplating a similar train of unfruitful motions, finds that at no time in the future will the equilibrium be disturbed. But where evolution is progressing this is no longer conceivable, as being contradictory to the very idea of progressive development. In this case Infinite Intelligence

necessarily finds, as the result of his contemplation, the aggregation of matter, and the consequences arising therefrom.

The negation of so primary a material property as gravitation to
these primitive motions of (or in) the ether, probably involves
the negation of many properties we find associated with matter.
Possibly the quality of inertia, equally primary, is involved
with that of gravitation, and we may suppose that these two
properties so intimately associated in determining the motions of
bodies in space were conferred upon the primitive motions as
crystallographic attraction and rigidity are first conferred upon
the solid growing from the supersaturated liquid. But in some
degree less speculative is the supposition that the new order of
motions involved the transformation of much energy into the form
of heat vibrations; so that the newly generated matter, like the
newly formed crystal, began its existence in a medium richly fed
with thermal radiant energy. We may consider that the thermal
conditions were such as would account for a primitive
dissociation of the elements. And, again, we recall how the
physicist finds his estimate of the energy involved in mere
gravitational aggregation inadequate to afford explanation of
past solar heat. It is supposable, on such a hypothesis as we
have been dwelling on, that the entire subsequent gravitational
condensation and conversion of material potential energy, dating
from the first formation of matter to the stage of star
formation

may be insignificant in amount compared with the conversion of
etherial energy attending the crystallizing out of matter from
the primitive motions. And thus possibly the conditions then
obtaining involved a progressively increasing complexity of
material structure the genesis of the elements, from an
infra-hydrogen possessing the simplest material configuration,
resulting ultimately in such self-luminous nebula as we yet see
in the heavens.

The late James Croll, in his _Stellar Evolution_, finds objections
to an eternal evolution, one of which is similar to the
"metaphysical" objection urged in this paper. His way out of the
difficulty is in the speculation that our stellar system
originated by the collision of two masses endowed with relative
motion, eternal in past duration, their meeting ushering in the
dawn of evolution. However, the state of aggregation here
assumed, from the known laws of matter and from analogy, calls
for explanation as probably the result of prior diffusion, when,
of course, the difficulty is only put back, not set at rest. Nor

do I think the primitive collision in harmony with the number of relatively stationary nebula visible in space.

The metaphysical objection is, I find, also urged by George Salmon, late Provost of Trinity College, in favour of the creation of the universe.—(_Sermons on Agnosticism_.)

A. Winchell, in _World Life_, says: "We have not

the slightest scientific grounds for assuming that matter existed in a certain condition from all eternity. The essential activity of the powers ascribed to it forbids the thought; for all that we know, and, indeed, as the _conclusion_ from all that we know, primal matter began its progressive changes on the morning of its existence."

Finally, in reference to the hypothesis of a unique determination of matter after eternal duration in the past, it may not be out of place to remind the reader of the complexity which modern research ascribes to the structure of the atom.

www.ingramcontent.com/pod-product-compliance
Ingram Content Group UK Ltd.
Pitfield, Milton Keynes, MK11 3LW, UK
UKHW040817280325
456847UK00003B/495